Hadoop 离线分析实战

聂　强　付　雯　主　编

武春岭　李俊翰　童世华
　　　　李清莲　蓝　菊　副主编

段　科　卢　山　参　编

北京理工大学出版社
BEIJING INSTITUTE OF TECHNOLOGY PRESS

内容简介

本书包含7个项目，项目一介绍数据管理技术的发展，项目二介绍 Hadoop 平台的搭建与安装，项目三介绍数据清洗技术，项目四介绍数据仓库 Hive 的使用，项目五介绍 Flume 的应用，项目六介绍海量数据传输工具 Sqoop，项目七介绍 Azkaban 调度器。

全书以典型案例贯穿，采用任务驱动方式逐步进行教学设计，结合大赛、职业技能证书展开编写工作，知识点由浅入深、覆盖面广，适合大数据技术相关专业教学使用，同时，对专业爱好者来说也是一本不错的入门级参考资料。

图书在版编目（CIP）数据

Hadoop 离线分析实战/聂强，付雯主编．--北京：北京理工大学出版社，2021.8（2024.7 重印）

ISBN 978-7-5682-9489-8

Ⅰ.①H…　Ⅱ.①聂…②付…　Ⅲ.①数据处理软件　Ⅳ.①TP274

中国版本图书馆 CIP 数据核字（2021）第 017709 号

责任编辑：王玲玲　　**文案编辑**：王玲玲
责任校对：刘亚男　　**责任印制**：施胜娟

出版发行／北京理工大学出版社有限责任公司
社　　址／北京市丰台区四合庄路6号
邮　　编／100070
电　　话／（010）68914026（教材售后服务热线）
（010）68944437（课件资源服务热线）
网　　址／http：//www.bitpress.com.cn

版 印 次／2024年7月第1版第6次印刷
印　　刷／唐山富达印务有限公司
开　　本／787 mm×1092 mm　1/16
印　　张／18.5
字　　数／425千字
定　　价／55.00元

Foreword 前言

党的二十大报告中提出，推动战略性新兴产业融合集群发展，构建新一代信息技术、人工智能、生物技术等一批新的增长引擎。

数字经济和实体经济融合逐步加深，未来的数据分析将对实时性和实效性要求越来越高。大数据、人工智能、云计算等技术的应用，加速了科技的发展，改善了人们的生活，同时产生了大量的数据信息，科学、合理的对数据进行管理和利用，能够为人们未来的生活和工作带来更大便利。利用 Hadoop 离线分析技术对数据进行管理，使数据资源优势得以充分发挥，让决策者迅速把握用户关键需求，并为用户的需求提供实时响应。

针对大数据技术的迅猛发展，本书结合实际应用案例，坚持问题导向，通过梳理数据管理技术发展过程，引入了大数据的相关概念，并对 Hadoop 的整体技术生态进行了介绍，结合国家职业教育大数据技术专业教学资源库项目，积极发挥教育资源优势，培养造就大批德才兼备的高素质人才，服务于加快建设网络强国、数字中国。

本书以 Hadoop 生态圈技术作为大数据离线分析的工具，采用项目驱动的编写方式，精心设计了 7 个项目引导学习，覆盖了 Hadoop 平台的搭建、安装技术、数据清洗的方法、数据仓库的应用及 Flume、Sqoop、Azkaban 的部署应用。通过简单项目开展项目部署，内容深入浅出。具体为：

项目一从对数据的认识、数据的使用、数据的管理入手，通过案例让读者认识 Hadoop 技术。

项目二详细介绍了 Hadoop 平台搭建的基本知识，并介绍了验证 Hadoop 平台的基本方法。

项目三介绍了 HDFS 分布式文件系统体系架构和文件操作、MapReduce 分布式计算系统基本运行框架、YARN 的平台资源调度方式。

项目四详细介绍了 Hive 的搭建、组件的新增工作，以及如何用 Hive 来存储数据并执行查询分析。

项目五介绍了 Flume 组件的安装及其运行机制，同时介绍了日志采集的流程和方法。

项目六介绍了海量数据传输的原理和方法，以及 Sqoop 组件的安装和数据的导入/导出。

项目七介绍了 Azkaban 调度器的安装部署、数据库的导入及工作流的构建等内容。

本书由重庆电子工程职业学院的聂强、付雯教授主编，重庆电子工程职业学院的武春岭、李俊翰、童世华、李清莲、蓝菊副主编，中山职业技术学院段科、重庆翰海睿智大数据科技有限公司总经理卢山参编，重庆电子工程职业学院江恒也参与了本书的编写工作。

本书中引入大量的具备极高操作性的可运行案例代码，进一步降低了读者的实操难度。

尽管我们尽了最大努力，但书中难免有不妥之处，欢迎各界专家和读者朋友们给予宝贵意见。

编　者

Contents 目录

项目一
认识数据管理

项目描述

数字化时代的到来，使得数据量越来越多，因此需要对这些数据进行有效、科学的管理。数据管理是通过软件技术来实施的特殊操作。数据的多样化使许多行业在数据管理方面的难度加大，这就促使企业在开展数据管理工作时，创新管理方法，从而合理、科学地利用数据进行项目开发和管理。

项目分析

通过本项目的学习，了解数据管理技术及 Hadoop 分布式平台架构技术。

任务 1　认识数据管理技术

1. 早期的数据管理技术

美国统计学家赫尔曼·霍尔瑞斯为了统计 1890 年的人口普查数据，发明了一台电动器来读取卡片上的洞数，这台设备使美国用了 1 年时间就完成了原本需要耗时 8 年的人口普查工作。这就是最早所说的数据管理技术，即通过大量的分类、比较和表格绘制的机器运行数百万穿孔卡片来进行数据的处理，其运行结果在纸上打印出来或制成新的穿孔卡片。此时的数据管理就是对穿孔卡片进行物理上的存储和处理。

2. 早期的数据库技术

最早出现的是网状数据库系统，1961 年美国通用电气公司开发出世界上第一个网状数据库系——继承数据存储（Integrated Data Store，IDS），奠定了网状数据库的基础，并在当时得到了广泛的应用。

层次数据库系统是紧随网状数据库系统出现的，最著名的层次数据库是 IBM 公司在 1968 年开发的 IMS（Information Management System）。由于网状数据库模型对于层次和非层次结构的事务都具备良好的兼容性，因此层次数据库系统不如网状数据库系统使用广泛。

网状数据库和层次数据库虽然已解决了数据集中和数据共享问题，但在数据独立性和数据抽象级别上仍存在较大不足。1970 年，IBM 研究员 E. F. Codd 博士发表了“A Relational Model of Data for Large Shared Data Banks”论文，提出关系模型的概念。关系型模型建立后，IBM 公司在 1979 年实现了 SQL 数据库管理系统。1976 年，霍尼韦尔公司开发了第一个商用关系数据库系统 Multics Relational Data Store（MRDS）。

经过长时间的发展，数据库技术越来越成熟、完善，代表产品有 Oracle 公司的 Oracle、MySQL，IBM 公司的 DB2、Informix，微软公司的 MS SQL Server 等。

3. 决策支持系统和数据仓库

决策支持系统是诞生于 20 世纪 60 年代，辅助决策者通过数据、模型和知识，以人机交互方式进行半结构化或非结构化决策的计算机应用系统，是管理信息系统向更高一级发展而产生的先进信息管理系统。其为决策者提供分析问题、建立模型、模拟决策过程和方案的环境，调用各种信息资源和分析工具，帮助决策者提高决策水平和质量。

1988 年，为解决企业集成问题，IBM 公司的研究员 Barry Devlin 和 Paul Murphy 提出了数据仓库（Data Warehouse）的概念。它是决策支持系统和联机分析应用数据源的结构化数据环境，是一个面向主题的（Subject Oriented）、集成的（Integrated）、相对稳定的（Relative Stable）、反映历史变化（Time Variant）的数据集合，用于支持管理决策（Decision Making Support），服务于数据管理。

4. 数据挖掘和商业智能

数据挖掘指通过分析大量的数据来展示数据之间隐藏的关系、模式和趋势，从而为决策者提供新的知识。这里“挖掘”用来比喻在海量数据中寻找有用知识，很困难，也很难得。

数据挖掘是数据量快速增长的直接产物。自从有了数据仓库，数据挖掘如虎添翼，在实业界不断产生化腐朽为神奇的故事。最脍炙人口的当属啤酒和尿布。Wal－Mart（沃尔玛）拥有世界上最大的数据仓库，在一次购物车分析之后，研究人员发现，跟尿布一起搭配购买最多的商品居然是啤酒。经过大量的跟踪调查，研究人员发现，在美国，一些年轻的父亲经常被妻子“派”到超市去购买婴儿尿布，有 30%～40% 的新生爸爸会顺便买点啤酒犒劳自己。沃尔玛随后对啤酒和尿布进行了捆绑销售，不出意料，销售量双双增加。

1989 年，高德纳 IT 咨询公司（Gartner Group）提出了商业智能的概念。商业智能（Business Intelligent，BI）即一系列以数据为支持、辅助商业决策的技术和方法。

商业智能是利用数据仓库、数据挖掘技术对客户数据进行系统的储存和管理，并通过各种数据统计、分析工具对客户数据进行分析，提供各种分析报告，为企业的各种经营活动提供决策信息。

5. 大数据

1980 年，著名未来学家阿尔文·托夫勒的《第三次浪潮》一书中，将大数据称为“第三次浪潮的华彩乐章”。但是在相当长的一段时间里，大数据技术并没有得到实质性的发展。

自 1995 年起，随着信息技术的发展，大量信息产生，随之出现了的结构化、非结构化数据，彻底打乱了传统数据管理技术、数据挖掘、商业智能的节奏。传统数据管理技术面临着前所未有的压力和挑战，数据挖掘领域和商业智能领域对于高效处理海量数据的技术需求极为迫切。以 Hadoop 为核心的大数据相关技术此刻飞速发展，大数据技术走向成熟。

2011 年，美国咨询公司麦肯锡发表了研究报告“Big data：The next frontier for innovation，competition，and productivity”，报告首次提出“大数据时代的到来”。

目前，被广泛引用的数据均来自国际数据公司（IDC）发布的研究报告“The Digital Universe of Opportunities：Rich Data and the Increasing Value of the Internet of Things”，该报告称：2013 年数字世界项目统计得出，全球数据总量为 4.4 ZB。同时预测，在 2020 年将达到 44 ZB。

研究机构 Gartner 对大数据给出的定义为：大数据是需要新处理模式才能具有更强的决策力、洞察发现力和流程优化能力的海量、高增长率和多样化的信息资产。

麦肯锡全球研究对大数据所给出的定义是：一种规模大到在获取、存储、管理、分析方面大大超出了传统数据库软件工具能力范围的数据集合，具有海量的数据规模（Volume）、快速的数据流转（Velocity）、多样的数据类型（Variety）和低的价值密度（Value）四大特征。

其实，大数据的真正意义在于如何高效地从这些价值密度较低的海量数据中寻找价值，为各行各业的商业决策服务。

任务 2　初识 Hadoop

谷歌的“三驾马车”

1.“三驾马车”与 Hadoop

谈到 Hadoop，不得不提谷歌公司的“三驾马车”：Google FS（GFS）、MapReduce、BigTable。

谷歌的“三驾马车”主要为谷歌的核心搜索业务服务。谷歌搜索要存储整个互联网的内容，同时要基于内容构建倒排索引。为了能够大幅提升计算效率并降低硬件成本，谷歌开发了三项技术：

- 谷歌文件系统（GFS），基于大量的、廉价的个人计算机构建分布式的海量的数据存储系统，可轻松存储整个互联网内容。
- 海量数据计算引擎（MapRecude），用来大规模地处理整个互联网的所有文档，建立倒排索引。虽然其有天然的缺陷，但是大幅提升了倒排索引的计算效率。
- 键值存储系统（BigTable），可以存储一个主键为不同时期的多个版本的值，通过使用互联网地址作为主键的方式，可以实现增量更新索引。

虽然谷歌公司没有公布这三个产品的源代码，但是发布了这三个产品的详细设计论文，从而奠定了 Hadoop 的基础。

2004 年，受到 Google 发布的 GFS 和 MapReduce 思想的启发，Doug Cutting 等人用两年的时间实现了 DFS 和 MapReduce 机制，使得 Nutch 性能大幅提升。

2005 年，Hadoop 作为 Lucene 的子项目 Nutch 的一部分正式引入 Apache 基金会。

2006 年，Hadoop 作为一套完整、独立的软件框架被分离，正式命名为 Hadoop，成为 Apache 顶级项目。

2008 年，Hadoop 赢得世界最快的 1 TB 数据排序记录（在 900 个节点上用时 209 s）。

2009 年，雅虎团队使用 Hadoop 对 1 TB 数据进行排序，只花费了 62 s。自此，Hadoop 作为大数据离线分析的开源框架在大数据领域显露峥嵘。

2. 认识 Hadoop

Hadoop 是一套分布式系统基础框架。用户可在不了解底层细节的前提下基于 Hadoop 框架运行分布式程序，而 Hadoop 框架将为分布式程序提供可靠性和数据处理能力。

Hadoop 实现了 MapReduce 编程模型，它能够将分布式应用程序自动地分成小的工作片段，每一个工作片段都可以在集群的任何节点执行。

Hadoop 还提供了一个分布式文件系统（HDFS），用户可以在计算节点存储数据。HDFS 容错性高，能够提供高吞吐量来访问应用程序的数据。

（1）Hadoop MapReduce

MapReduce 是一种编程模型，它通过使用键值对数据集的分布式操作序列实现大型分布式计算。在 Map 阶段，框架将输入数据集分拆为大量的片段，将每个片段分配给各 Map 任务。为运行在 Map 任务的所有集群节点分发大量的 Map 任务，每个 Map 任务都从框架为其分配的数据集片段中获取键值对数据，经过计算后生成新的键值对数据。Map 任务将调用用户定义的 Map 函数完成每一个键值对数据的转化，从而转化为新的键值对数据。

Map 阶段结束后，框架将键值对元组集拆分为与 Reduce 任务数量相同的片段。

在 Reduce 阶段，每个 Reduce 任务都会从框架为其分配的数据集片段中获取键值对元组片段作为输入数据，调用用户定义的 Reduce 函数进行计算，完成将元组转化为输出键值对数据。与 Map 阶段机制相同，Reduce 阶段框架也会为运行 Reduce 任务的所有集群节点分发大量的 Reduce 任务，并将键值对元组数据片段发送给每个 Reduce 任务。

（2）Hadoop DFS

Hadoop DFS（HDFS）组件是一个硬件设备集群，用于存储海量数据文件，并且从机制上能够保证数据存储的可靠性。其设计灵感来自 GFS。HDFS 将大文件以数据块的方式分割存储，除了最后一个数据块外，所有的数据块具有相同的大小。为了实现高容错，同一数据文件的数据块被创建为多个副本进行分布式存储，框架提供了配置文件，便于用户根据实际业务需求配置数据块的大小及副本数量。HDFS 中的文件采用 “WORM”（write once read many）机制，数据是只读的，不允许对数据进行修改。

HADOOP DFS 简介

3. Hadoop 技术生态组件

以 MapReduce 和 HDFS 为核心的 Hadoop 软件架构，由于其高可靠性、高效及可伸缩的特点，使 Hadoop 开源社区活跃。围绕着 Hadoop 用于解决特定问题的一系列开源组件不断涌现，构建了一套以 Hadoop 为核心的技术生态。其中，使用最为广泛、最重要的几个组件如下。

（1）YARN

Apache Hadoop YARN（Yet Another Resource Negotiator）是一种全新的 Hadoop 资源管理器，是一个通用资源管理系统和调度平台，可以为层应用提供统一的资源管理和调度，它的引入为集群在利用率、资源统一管理和数据共享等方面带来了巨大好处。

第一代 MapReduce（称为 MRv1）是目前使用的标准大数据处理系统，但是在超大型集群上，其架构缺陷就被暴露出来了。当集群包含的节点超过 4 000 个时，就会表现出一定的不可预测性，其中最大的问题是级联故障，由于框架要尝试复制数据和重载活动的节点，所以任何一个故障都可能通过网络泛化导致整个集群的运行状况严重恶化。

YARN 分层结构的本质是 ResourceManager。它控制整个集群并管理应用程序向基础计算资源的分配。ResourceManager 将各个资源部分（计算、内存、带宽等）精心安排给基础 NodeManager（YARN 的每节点代理）。ResourceManager 还与 ApplicationMaster 一起分配资源，与 NodeManager 一起启动和监视它们的基础应用程序。

随着 YARN（称为 MRv2）的出现，开发者不再受 MapReduce 开发模式约束，而是可以创建更复杂的分布式应用程序。实际上，可以把 MapReduce 模型看成 YARN 架构可运行的一些应用程序中的一个子程序，只是为自定义开发公开了基础框架的更多功能。YARN 的使用模型几乎没有限制，不再需要与一个集群上可能存在的其他更复杂的分布式应用程序框架相隔离。随着 YARN 变得更加健全，它有能力取代其他分布式处理框架，从而完全消除专用于其他框架的资源开销，简化整个系统。

（2）Hive

The Apache Hive 是一个基于 Hadoop 的数据仓库基础架构，提供了类 SQL 查询语句（HQL）完成海量数据的汇总、查询和分析。

Hive 起源于 Facebook。Facebook 每天要产生大量的社交网络数据，通过 MapReduce 编程虽能实现对海量数据的分析，但业务逻辑变更导致的代码变更所产生的成本极大。因此，Facebook 研发并开源了 Hive，用于解决海量结构化日志数据的统计问题。

Hive 能够将存储在 HDFS 上的结构化数据文件映射为一张数据库表，通过简单的类 SQL 查询语句完成对海量数据的数据查询、统计的功能。Hive 可以将类 SQL 语句转化为 MapReduce 作业在分布式集群环境运行，学习成本低，适合基于数据仓库进行统计分析。

（3）ZooKeeper

Apache ZooKeeper 是一个分布式协调服务，用于维护配置信息、命名、提供分布式同步服务和组服务。

ZooKeeper 最早起源于雅虎研究院的一个研究小组。工作人员发现，雅虎内部的很多大型系统都需要依赖类似的系统来进行分布式协调，但这些系统往往都存在单点分布式问题。雅虎的研发人员试图开发一个通用的无单点问题的分布式协调框架，使开发人员能够将精力集中在业务处理逻辑上。

ZooKeeper 的设计目标是将那些杂且易错的分布式一致性服务封装起来，构建一个高效、可靠的原语集，并将一系列简单、易用的接口提供给用户使用。ZooKeeper 是一个典型的分布式数据一致性解决方案，分布式应用程序可以基于 ZooKeeper 实现诸如数据发布/订阅、负载均衡、命名服务、分布式协调/通知、集群管理、Master 选举、分布式锁和分布式队列等功能。

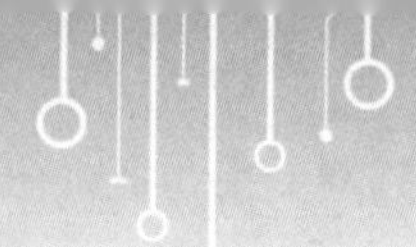

（4） Flume

Apache Flume 是一个高可用的、高可靠的、分布式的海量日志采集、聚合和传输的系统。其是基于流式数据的简单、灵活的框架。同时，它提供简单的可扩展数据模型来支持在线分析应用程序。

Flume 最早的版本是由 Cloudera 发布的，称为 Flume OG，其能够实时地将分布在不同节点、设备上的日志搜集到 HDFS。但随着 Flume 功能的扩展，Flume OG 代码工程臃肿、核心组件设计不合理、核心配置不标准等一系列缺陷逐渐暴露。

2011 年 10 月 22 日，Cloudera 完成了 Flume－782，对 Flume 进行了改动：重构核心组件、核心配置及代码架构，重构后的版本统称为 Flume NG。改版后的 Flume NG 正式纳入 Apache 旗下，改名为 Apache Flume。

Flume 的核心是把数据从数据源（source）收集过来，再将收集到的数据送到指定的目的地（sink）。为保证数据传输的成功率，在送到目的地（sink）之前，会将数据先缓存起来（channel），待数据真正到达目的地（sink）后，Flume 才会删除缓存的数据。

在整个数据传输的过程中，流动的是 event，保证事务是在 event 的级别。event 是 Flume 传输数据的基本单元，如果是文本文件，通常是一条记录，event 也是事务的基本单位。event 从 source 流向 channel，再到 sink，其本身是一个字节数组，可携带头信息（headers）。event 代表着一个数据的最小完整单元，从外部数据源来，向外部目的地去。

Flume 的数据源是可定制的，可以用于传输大量事件数据，包括但不限于网络流量数据、社交媒体生成的数据、电子邮件消息和几乎所有可能的数据源。

（5） Sqoop

Apache Sqoop 是用于在 Hadoop 和结构化存储（例如关系型数据库）之间高效、批量传输数据的工具。用户可使用 Sqoop 将存储在关系型数据库（如 MySQL、MS SQL Server、Oracle 等）中的数据导入到 HDFS、Hive 或 HBase 中，也可以将存储在 HDFS、Hive 或 HBase 中的数据提取并导出到结构化数据库中。

4. Hadoop 发行版

由于 Hadoop 遵从 Apache 开源协议，任何人可以对其进行修改，并作为开源或商业产品发布/销售，因此，市面上出现了很多 Hadoop 版本，有 Intel 发行版、华为发行版、Cloudera 发行版（CDH）、Hortonworks 版本、MapR 的 MapR 产品等。

（1） Apache Hadoop

Apache Hadoop 是 Apache 官方的社区版本，是最原始的版本，所有发行版均是以此为基础进行改进的。

其发行版本主要有三代：

- Hadoop 1. x

  第一代包含三个版本，分别是 0. 20. x、0. 21. x 和 0. 22. x。其中，0. 20. x 最后演化成 1. 0. x，变成了稳定版；而 0. 21. x 和 0. 22. x 则支持 NameNode HA 等新的重大特性。

- Hadoop 2. x

  第二代 Hadoop 包含两个版本，分别是 0. 23. x 和 2. x，它们是一套全新的架构，均包

含 HDFS Federation 和 YARN 两个系统。

■ Hadoop 3. x

Hadoop 2. 0 是基于 JDK 1. 7 开发的，而 JDK 1. 7 在 2015 年 4 月已停止更新，这直接迫使 Hadoop 社区基于 JDK 1. 8 重新发布一个新的 Hadoop 版本，即 Hadoop3. x。Hadoop 3. 0 中引入了一些重要的功能和优化，包括 HDFS 可擦除编码、多 NameNode 支持、MR Native Task 优化、YARN 基于 cgroup 的内存和磁盘 I/O 隔离、YARN container resizing 等。

（2） Cloudera Hadoop

Cloudera Hadoop 是由 Cloudera 维护的 Hadoop 发行版，通常将该发行版称为 CDH（Cloudera Distribution Hadoop）。CDH Hadoop 版本的划分非常清晰，并且经过了严格的测试环节。CDH 包括收费的企业版及完全开源的社区版。

截至目前，CDH 共发行 5 个版本，其中，前三个版本已不再更新，最新的两个版本分别是 CDH4 和 CDH5。CDH4 基于 Hadoop 2. 0，CDH5 则基于 Hadoop 2. 2、Hadoop 2. 3、Hadoop 2. 5、Hadoop 2. 6，并且还在不断地更新。相比于 Apache Hadoop，CDH 发行版在兼容性、安全性和稳定性上有较大的增强。

（3） Hortonworks Data Platform

Hortonworks Data Platform 是由美国大数据公司 Hortonworks 开发的企业级 Hadoop 平台，通常将该发行版称为 HDP。HDP 完全是在开源的环境下设计、开发和构建的，它以 YARN 作为其架构中心，该平台支持一系列处理方法——批处理、交互式处理、实时处理。HDP 的功能包括数据管理、数据访问、数据管制与集成、运营、安全性。

（4） Hadoop 版本的选择方法

由于 Hadoop 版本比较多，很多用户不知怎样选择。实际上，当前 Hadoop 共有三个版本：Hadoop 1. x、Hadoop 2. x 和 Hadoop 3. x。其中，Hadoop 1. 0 由一个分布式文件系统 HDFS 和一个离线计算框架 MapReduce 组成，而 Hadoop 2. 0 则包含一个支持 NameNode 横向扩展的 HDFS、一个资源管理系统 YARN 和一个运行在 YARN 上的离线计算框架MapReduce。Hadoop 3. x 尚且不够成熟和稳定。

当决定是否将某个软件用于开源环境时，通常需要考虑以下几个因素：

①是否为开源软件，即是否免费。

②是否有稳定版，一般官网会有说明。

③是否经实践验证，这个可以通过查看是否有企业应用案例体现。

④是否有强大的社区支持，当遇到问题时，是否能够通过社区、论坛等资源获取解决问题的办法。

学习笔记

项目二 Hadoop平台的搭建与安装

项目描述

假设需要完成一个大数据离线分析项目，从现在开始，进入Hadoop离线分析平台的学习。

项目分析

以“大数据相关技术构建市场招聘需求监控分析系统”为例，对项目进行分析，组建团队，分析项目需求，进行项目设计、开发。这里从项目团队组建开始。

假设将自己的角色定义为一名系统运维组人员，需要通过不断的学习和训练来完成项目组分配给自己的工作。

通常情况下，企业在构建大数据分析项目团队时，会采取两种组织方式：

1. 项目型团队

采用该方式组织的团队包括项目组的全部职能岗位，需要项目经理、系统架构师、数据采集工程师、数据清洗工程师、数据分析工程师、数据可视化工程师及大数据测试工程师等诸多岗位，这些岗位通常统一由项目经理对其进行管理。

2. 职能型团队

采用该方式组织的团队不设项目组，通常会包括一位项目经理及一位系统设计师。由系统设计师根据项目需求对项目整体进行模块化设计与拆分，项目经理按照软件开发工期进行项目安排，而实际的工作内容都是由各职能部门或职能小组完成的。项目组里同样需要系统运维组、数据采集组、数据清洗组、数据分析组、数据可视化开发组及软件测试组分别承担各自职能范围内的工作任务，项目经理不能管理各职能组的内部人员。

本项目以职能型团队的组织方式为例。

经过需求分析工程师、软件设计工程师的工作，已经对系统进行了整体的设计，系统由数据采集子系统、数据存储与分析平台、数据可视化子系统三个子系统构成，其中数据采集子系统、MapReduce、数据分析、数据可视化子系统为定制开发内容，其他组件均采用主流大数据开源组件。系统结构如图 2－1 所示。

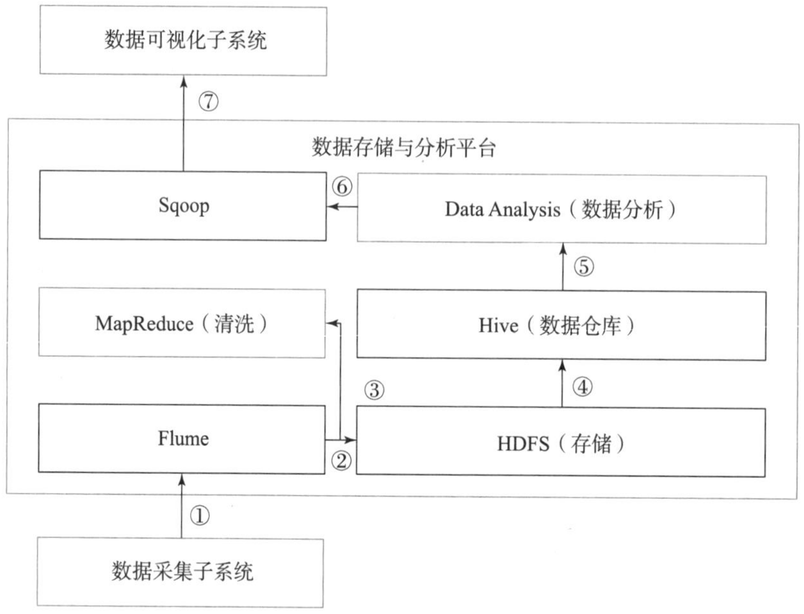

图 2－1 系统结构

作为系统运维组成员，项目经理分配给我们的工作有如下几项：

任务 1：搭建 Hadoop 2.6.0 基础开发环境，为数据清洗提供支撑平台。

任务 2：在开发环境中增加 Hive 1.1.0 组件，为数据分析提供支撑平台。

任务 3：在开发环境中增加 Flume 1.6.0 组件，实现与数据采集的集成。

任务 4：在开发环境中增加 Sqoop 1.4.7 组件，实现与数据可视化的集成。

任务 5：搭建工作流任务调度系统 Azkaban，实现自动化运维。

根据项目经理安排的工作，从第一项工作开始，搭建 Hadoop 2.6.0 基础开发环境，为数据清洗提供支撑平台。

任务 1　Hadoop 平台搭建基础

HADOOP 的三种部署模式

完整的 Apache Hadoop 2.x 发行版包含以下四个组件：

- Hadoop Common：为其他 Hadoop 组件提供基础配置。

- Hadoop Distributed File System：可靠的、高吞吐的分布式文件系统。
- Hadoop MapReduce：分布式离线并行计算框架。
- Hadoop YARN：集群资源统一管理及调度组件。

Hadoop 支持三种部署模式：

- 单机模式

  默认情况下，Hadoop 将被配置为以非分布式模式运行一个独立的 Java 进程。在此情况下，所有程序都运行在同一个 JVM 上，不需要任何守护进程来保证其可靠性。该模式下可以高效地对 MapReduce 程序进行调试。
- 伪分布模式

  Hadoop 将在一个节点上以伪分布模式运行，每一个 Hadoop 守护进程会作为一个独立的 Java 运行。伪分布模式可以认为是一种只有一个节点的分布模式，可以部分模拟一个极小规模的 Hadoop 集群。
- 完全分布模式

  这是一种具有实际意义的 Hadoop 运行模式，运行的 Hadoop 能够充分发挥其分布式计算、分布式存储及资源调度的全部优势。所有的企业项目都是采用完全分布模式进行离线分析平台的搭建的。

1. 运行平台支持

MOBAX SSH 工具配置

理论上讲，Hadoop 在 Linux、Windows、Mac OS、UNIX 操作系统上都能运行，但从 Apache Hadoop 的官方描述上讲，GUN/Linux 是 Hadoop 产品的开发和运行平台，并且 Hadoop 已经在由具有 2 000 个节点的 GUN/Linux 主机组成的集群系统上得到验证。因此，Linux 操作系统是运行 Hadoop 的最稳定且应用最广泛的操作系统平台。

2. 所需软件

由于 Hadoop 使用 Java 作为产品开发语言，因此，无论使用哪个版本的操作系统，都需要安装 Java。

（1）Java

Hadoop 2. x 对 Java 6、Java 7、Java 8 都能进行良好的支持。但 Oracle 公司停止了 Java 6、Java 7 的更新，Hadoop 开发者社区中的开发者推荐使用 Java 8，基于此，选择 Oracle 发布的 Java 8 作为平台的运行基础。

（2）SSH

Hadoop 控制脚本依赖 SSH 来执行针对整个集群的操作，为支持无缝式工作，需要在集群内的全部主机安装 SSH 并且保证 SSHD 进程一直运行。之外，配置主机之间的免密登录能够更好地保证 Hadoop 运行过程中不需要大量的人工介入。

3. Hadoop 集群搭建流程

搭建一个 Hadoop 完全分布式集群需要有以下几个步骤：

(1) Hadoop 集群规划

环境搭建前，需要根据硬件条件及项目需求对集群进行整体规划，通常包括主机规划、软件规划、网络拓扑结构规划及集群规划。

(2) 运行平台搭建

按照集群规划方案进行安装主机操作系统、修改主机名称、配置主机 IP 地址等工作。

(3) 安装配置支持软件

按照集群规划方案配置集群间的免密登录，为每台主机安装 JDK；为确保 SSHD 能够开机运行，还需要额外做一些确认工作。

(4) 安装配置 Hadoop

按照集群规划方案为每台主机安装 Hadoop，并将 Hadoop 配置为与集群规划方案一致的角色，同时，需要完成各主机的 Hadoop 相关进程的初始化及启动工作。

(5) 启动 Hadoop 并验证

这时 Hadoop 已经按照集群规划完成了全部的安装部署。为确保安装配置的 Hadoop 能够正常运行，需要重新启动 Hadoop 的各个进程、查看各节点的状态、访问 Web UI，以确定其运行正常。运行一个 Hadoop 内置的 WordCount 验证程序，验证 Hadoop 的功能能够正常运行。

任务 2　Hadoop 集群规划 

从本任务开始，将了解如何安装、配置和管理 Hadoop 完全分布式集群。首先，对即将部署的 Hadoop 集群进行总体规划。

1. 主机规划

主机规划见表 2-1。

表 2-1　主机规划

主机名	IP 地址	CPU 核心数	内存/GB	硬盘容量/GB
Master	192.168.3.190	8 核 64 位 (不少于 4 核)	16	500 以上
Slave1	192.168.3.191	6 核 64 位 (不少于双核)	8	500 以上
Slave2	192.168.3.192	6 核 64 位 (不少于双核)	8	500 以上

2. 软件规划

软件规划见表 2-2。

表 2-2　软件规划

软件	版本号	位数	版本说明
操作系统	CentOS 7.4.1708 Mini	64 位	
JDK	1.8.X	64 位	
Hadoop	2.6.0	—	官方稳定版

除非特别说明，本书中所有的系统操作及命令均运行在 CentOS 7 操作系统上，其他 Linux 操作系统的命令与 CentOS 的可能略有不同。

3. 网络拓扑结构规划

网络拓扑结构如图 2-2 所示。

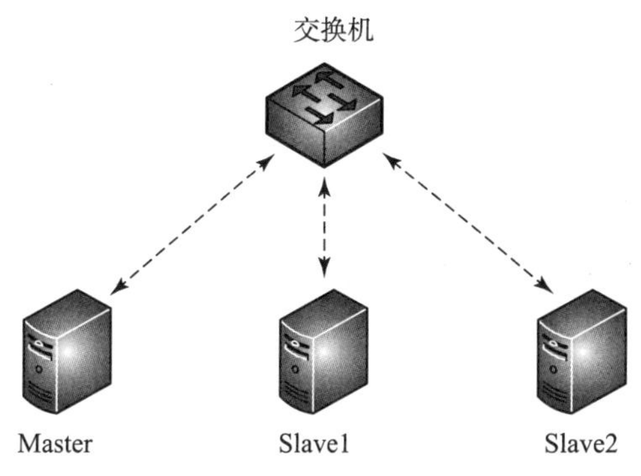

图 2-2　网络拓扑结构

4. 集群规划

集群规划见表 2-3。

表 2-3　集群规划

角色	Master	Superintendent1	Superintendent2
NameNode	●	×	×
DataNode	●	●	●
ResourceManager	●	×	×
NodeManager	×	●	●

■ NameNode

NameNode 管理 HDFS 文件系统的命名空间，它维护着文件系统及文件系统内所有的文件和目录，这些信息以两个文件形式永久保存在本地磁盘上：命名空间镜像文件和编辑日志文件。NameNode 也记录着每个文件中各个数据块所在的数据节点信息，但它并不永久保存数据块的位置信息，因为这些信息会在系统启动时根据数据节点信息重建。

■ DataNode

DataNode 是 HDFS 文件系统的工作节点，它们根据需要存储并检索数据块（受 NameNode 调度），并且定时向 NameNode 发送它们存储的数据块的列表。

■ ResourceManager

ResourceManager 是管理集群所有可用资源的中心节点，并能够帮助管理 YARN 上的分配部署 applications。它和每个节点上的 NodeManagers（NMs）及 ApplicationMasters（AMs）一起工作。

■ NodeManager

NodeManager 是 YARN 中每个节点上的代理，它管理 Hadoop 集群中单个计算节点，包括与 ResourceManger 保持通信，监督 Container 的生命周期管理，监控每个 Container 的资源使用（内存、CPU 等）情况，追踪节点健康状况，管理日志和不同应用程序用到的附属服务（auxiliary service）。

任务3　运行平台搭建

默认三台主机已经按照规划要求完成了操作系统的安装，并且已经按照网络拓扑结构规划要求进行了网络连接。平台准备阶段需要完成如下操作：

修改主机名称

1. 修改主机名称

为方便对集群主机进行管理，需要对集群主机进行身份确认，以便区别该主机节点的身份。操作系统安装完毕后，所有节点默认的主机名称都是 localhost. localdomain，按照集群规划的要求，修改各节点的主机名称。

（1）修改 Master 主机名

使用 root 用户登录 Master 主机，输入以下命令修改 Master 主机名称：

```
1.[root@192-168-3-190 ~]# sudo vi /etc/hostname
2.YYR
```

（2）修改 Superintendent1 主机名

使用 root 用户登录 Superintendent1 主机，输入以下命令修改 Superintendent1 主机名称：

```
1.[root@192-168-3-191 ~]# sudo vi /etc/hostname
2.Superintendent1
```

（3）修改 Superintendent2 主机名

使用 root 用户登录 Superintendent2 主机，输入以下命令修改 Superintendent2 主机名称：

```
1.[root@192-168-3-192 ~]# sudo vi /etc/hostname
2.Superintendent2
```

配置完成后，重启主机（reboot 命令），使用 root 用户登录系统后，查看主机名称，验证是否修改成功。当主机名称显示如下时，表示修改成功：

```
[root@YYR ~]#

[root@Superintendent1 ~]#

[root@Superintendent2 ~]#
```

2. 配置 IP 地址

为保证 Hadoop 集群稳定运行，为集群内的每一台主机节点配置唯一的静态 IP 地址，按照集群规划的要求，为主机配置静态 IP 地址。

（1）配置 Master 主机 IP 地址

使用 root 用户登录 Master 主机，输入以下命令来修改 Master 主机 IP 地址：

配置主机 IP 地址

```
[root@YYR ~]# sudo vi /etc/sysconfig/network-scripts/ifcfg-eth0
TYPE=Ethernet
PROXY_METHOD=none
BROWSER_ONLY=no
BOOTPROTO=dhcp
DEFROUTE=yes
IPV4_FAILURE_FATAL=no
IPV6INIT=yes
IPV6_AUTOCONF=yes
IPV6_DEFROUTE=yes
IPV6_FAILURE_FATAL=no
IPV6_ADDR_GEN_MODE=stable-privacy
NAME=eth0
UUID=1ae0279a-fc88-4dd5-82ba-d197f90df186
DEVICE=eth0
ONBOOT=yes

# 以下为新增加的静态 IP 地址配置项
# 静态 IP 地址
IPADDR=192.168.3.190
# 子网掩码(需要根据实际情况配置)
NETMASK=255.255.255.0
# DNS(需要根据实际情况配置)
DNS=202.106.0.20
```

```
# 重启网络服务,使得网络配置生效
[root@YYR ~]# sudo service network restart
```

（2）配置 Superintendent1 主机 IP 地址

使用 root 用户登录 Superintendent1 主机，输入以下命令来修改 Superintendent1 主机 IP 地址：

```
[root@Superintendent1 ~]# sudo vi /etc/sysconfig/network-scripts/ifcfg-eth0
TYPE=Ethernet
PROXY_METHOD=none
BROWSER_ONLY=no
BOOTPROTO=dhcp
DEFROUTE=yes
IPV4_FAILURE_FATAL=no
IPV6INIT=yes
IPV6_AUTOCONF=yes
IPV6_DEFROUTE=yes
IPV6_FAILURE_FATAL=no
IPV6_ADDR_GEN_MODE=stable-privacy
NAME=eth0
UUID=1ae0279a-fc88-4dd5-82ba-d197f90df186
DEVICE=eth0
ONBOOT=yes
# 以下为新增加的静态 IP 地址配置项
# 静态 IP 地址
IPADDR=192.168.3.191
# 子网掩码(需要根据实际情况配置)
NETMASK=255.255.255.0
# DNS(需要根据实际情况配置)
DNS=202.106.0.20
# 重启网络服务,使得网络配置生效
[root@Superintendent1 ~]# sudo service network restart
```

（3）配置 Superintendent2 主机 IP 地址

使用 root 用户登录 Superintendent2 主机，输入以下命令来修改 Superintendent2 主机 IP 地址：

```
[root@Superintendent2 ~]# sudo vi/etc/sysconfig/network-scripts/ifcfg-eth0
TYPE=Ethernet
PROXY_METHOD=none
BROWSER_ONLY=no
BOOTPROTO=dhcp
DEFROUTE=yes
IPV4_FAILURE_FATAL=no
IPV6INIT=yes
IPV6_AUTOCONF=yes
IPV6_DEFROUTE=yes
IPV6_FAILURE_FATAL=no
IPV6_ADDR_GEN_MODE=stable-privacy
NAME=eth0
UUID=1ae0279a-fc88-4dd5-82ba-d197f90df186
DEVICE=eth0
ONBOOT=yes

#以下为新增加的静态IP地址配置项
#静态IP地址
IPADDR=192.168.3.192
#子网掩码(需要根据实际情况配置)
NETMASK=255.255.255.0
#DNS(需要根据实际情况配置)
DNS=202.106.0.20

#重启网络服务,使得网络配置生效
[root@Superintendent2 ~]# sudo service network restart
```

修改完成后，输入以下命令来验证各主机节点IP地址是否正确：

```
[root@YYR ~]# ip addr
1:lo: <LOOPBACK,UP,LOWER_UP> mtu 65536 qdisc noqueue state UNKNOWN qlen 1
    link/loopback 00:00:00:00:00:00 brd 00:00:00:00:00:00
    inet 127.0.0.1/8 scope host lo
       valid_lft forever preferred_lft forever
    inet6 ::1/128 scope host
```

```
       valid_lft forever preferred_lft forever
2: eth0: <BROADCAST,MULTICAST,UP,LOWER_UP> mtu 1500 qdisc pfifo_fast state UP qlen 1000
    link/ether fa:b0:76:cf:10:00 brd ff:ff:ff:ff:ff:ff
    inet 192.168.3.190/24 brd 192.168.3.255 scope global eth0
       valid_lft forever preferred_lft forever
    inet6 fe80::61f8:251f:d743:b358/64 scope link
       valid_lft forever preferred_lft forever
```

```
[root@Superintendent1 ~]# ip addr
1: lo: <LOOPBACK,UP,LOWER_UP> mtu 65536 qdisc noqueue state UNKNOWN qlen 1
    link/loopback 00:00:00:00:00:00 brd 00:00:00:00:00:00
    inet 127.0.0.1/8 scope host lo
       valid_lft forever preferred_lft forever
    inet6 ::1/128 scope host
       valid_lft forever preferred_lft forever
2: eth0: <BROADCAST,MULTICAST,UP,LOWER_UP> mtu 1500 qdisc pfifo_fast state UP qlen 1000
    link/ether fa:ff:a2:75:65:00 brd ff:ff:ff:ff:ff:ff
    inet 192.168.3.191/24 brd 192.168.3.255 scope global eth0
       valid_lft forever preferred_lft forever
    inet6 fe80::b663:fa95:4366:225c/64 scope link
       valid_lft forever preferred_lft forever
```

```
[root@Superintendent2 ~]# ip addr
1: lo: <LOOPBACK,UP,LOWER_UP> mtu 65536 qdisc noqueue state UNKNOWN qlen 1
    link/loopback 00:00:00:00:00:00 brd 00:00:00:00:00:00
    inet 127.0.0.1/8 scope host lo
       valid_lft forever preferred_lft forever
    inet6 ::1/128 scope host
       valid_lft forever preferred_lft forever
2: eth0: <BROADCAST,MULTICAST,UP,LOWER_UP> mtu 1500 qdisc pfifo_fast state UP qlen 1000
    link/ether fa:86:97:12:53:00 brd ff:ff:ff:ff:ff:ff
    inet 192.168.3.192/24 brd 192.168.3.255 scope global eth0
       valid_lft forever preferred_lft forever
```

```
    inet6 fe80::502c:2380:97a7:9d2c/64 scope link
       valid_lft forever preferred_lft forever
```

3. 配置域名解析规则

配置域名解析规则

修改了各主机节点名称及 IP 地址后，需要配置各个主机节点的域名解析规则，以保证集群各主机节点能够正确地解析所有节点的主机名称。

（1）配置 Master 主机节点域名解析规则

使用 root 用户登录 Master 主机，输入以下命令来配置 Master 主机域名解析规则：

```
[root@YYR ~]# sudo vi /etc/hosts
127.0.0.1   localhost localhost.localdomain localhost4 localhost4.localdomain4
::1         localhost localhost.localdomain localhost6 localhost6.localdomain6

# 以下为新增加项，IP 地址与主机名之间为半角空格
192.168.3.190 YYR
192.168.3.191 Superintendent1
192.168.3.192 Superintendent2
```

（2）配置 Superintendent1 主机节点域名解析规则

使用 root 用户登录 Superintendent1 主机，输入以下命令来配置 Superintendent1 主机域名解析规则：

```
[root@Superintendent1 ~]# sudo vi /etc/hosts
127.0.0.1   localhost localhost.localdomain localhost4 localhost4.localdomain4
::1         localhost localhost.localdomain localhost6 localhost6.localdomain6

# 以下为新增加项，IP 地址与主机名之间为半角空格
192.168.3.190 YYR
192.168.3.191 Superintendent1
192.168.3.192 Superintendent2
```

（3）配置 Superintendent2 主机节点域名解析规则

使用 root 用户登录 Superintendent2 主机，输入以下命令来配置 Superintendent2 主机域名解析规则：

```
[root@Superintendent2 ~]# sudo vi /etc/hosts
```

```
127.0.0.1   localhost localhost.localdomain localhost4 localhost4.localdomain4
::1         localhost localhost.localdomain localhost6 localhost6.localdomain6

# 以下为新增加项,IP 地址与主机名之间为半角空格
192.168.3.190 YYR
192.168.3.191 Superintendent1
192.168.3.192 Superintendent2
```

在各主机节点输入以下命令，以确认域名解析规则配置正确且已经生效：

```
# YYR 节点输入以下命令
[root@YYR ~]# ping Superintendent1
64 bytes from Superintendent1 (192.168.3.191):icmp_seq =1 ttl =64 time =0.409 ms
[root@YYR ~]# ping Superintendent2
64 bytes from Superintendent2 (192.168.3.192):icmp_seq =1 ttl =64 time =0.278 ms

# Superintendent1 节点输入以下命令
[root@Superintendent1 ~]# ping YYR
64 bytes from YYR (192.168.3.190): icmp_seq = 1 ttl = 64 time = 0.246 ms
[root@Superintendent1 ~]# ping Superintendent2
64 bytes from Superintendent2 (192.168.3.192):icmp_seq =1 ttl =64 time =0.262 ms

# Superintendent2 节点输入以下命令
[root@Superintendent2 ~]# ping YYR
64 bytes from YYR (192.168.3.190): icmp_seq = 1 ttl = 64 time = 0.240 ms
[root@Superintendent2 ~]# ping Superintendent1
64 bytes from Superintendent1 (192.168.3.191):icmp_seq =1 ttl =64 time =0.287 ms
```

4. 配置防火墙

Hadoop 集群主机的安全策略配置是一个很重要的环节，为了简化防火墙的配置工作，在本次工作任务中，将集群内全部主机节点的防火墙关闭，使 Hadoop 集群各节点能够相互

进行远程访问。在实际工作中，不建议直接关闭各个主机节点的防火墙，推荐通过防火墙安全策略的配置来实现主机节点之间的远程访问。

由于操作相同，以 YYR 操作为例说明操作过程，Superintendent1、Superintendent2 需要执行与 YYR 相同的操作。

```
# 关闭 selinux
[root@YYR ~]# sudo vi /etc/sysconfig/selinux

# This file controls the state of SELinux on the system.
# SELINUX=can take one of these three values:
#     enforcing - SELinux security policy is enforced.
#     permissive - SELinux prints warnings instead of enforcing.
#     disabled - No SELinux policy is loaded.

# 将该项设置为 disabled
SELINUX=disabled
# SELINUXTYPE=can take one of three two values:
#     targeted - Targeted processes are protected,
#     minimum - Modification of targeted policy. Only selected processes are protected.
#     mls - Multi Level Security protection.
SELINUXTYPE=targeted

# 禁止防火墙服务开机启动
[root@YYR ~]# sudo systemctl disable firewalld.service
Removed symlink /etc/systemd/system/multi-user.target.wants/firewalld.service.
Removed symlink /etc/systemd/system/dbus-org.fedoraproject.FirewallD1.service.

# 停止防火墙服务
[root@YYR ~]# sudo systemctl stop firewalld.service

# 查看防火墙服务状态
[root@YYR ~]# sudo systemctl status firewalld.service
firewalld.service - firewalld - dynamic firewall daemon
   Loaded: loaded (/usr/lib/systemd/system/firewalld.service; disabled; vendor preset: enabled)
```

```
   Active:inactive (dead)
     Docs:man:firewalld(1)

1 月 10 11:08:21 YYR systemd[1]: Starting firewalld - dynamic firewall daemon...
1 月 10 11:08:23 YYR systemd[1]:Started firewalld - dynamic firewall daemon.
1 月 10 11:08:24 YYR firewalld[748]:WARNING:ICMP type 'beyond - scope' is not supported b...v6.
1 月 10 11:08:24 YYR firewalld[748]:WARNING:beyond - scope:INVALID_ICMPTYPE:No supporte...me.
1 月 10 11:08:24 YYR firewalld[748]:WARNING:ICMP type 'failed - policy' is not supported ...v6.
1 月 10 11:08:24 YYR firewalld[748]:WARNING:failed - policy:INVALID_ICMPTYPE:No support...me.
1 月 10 11:08:24 YYR firewalld[748]:WARNING:ICMP type 'reject - route' is not supported b...v6.
1 月 10 11:08:24 YYR firewalld[748]:WARNING:reject - route:INVALID_ICMPTYPE:No supporte...me.
1 月 10 11:52:02 YYR systemd[1]: Stopping firewalld - dynamic firewall daemon...
1 月 10 11:52:03 YYR systemd[1]:Stopped firewalld - dynamic firewall daemon.
Hint:Some lines were ellipsized,use -l to show in full.

# 重启主机,使配置生效
[root@YYR ~]# reboot
```

任务 4　安装配置支持软件

完成运行平台搭建工作后，需要安装 Hadoop 运行支持软件，接下来完成 Java 和 SSH。如果集群环境无法连接互联网环境，需要提前准备好软件安装包。

1. 配置免密登录

OpenSSH（Open Secure Shell，开放安全 Shell）是使用 SSH，通过计算机网络加密通信的开源实现。它是取代由 SSH Communications Security 所提供的商用版本的开放源代码方案。目前 OpenSSH 是 OpenBSD 的子计划。

首先需要确保集群内所有主机节点都安装了 SSH，然后开始配置，以实现集群内主机之间的免密登录。

（1）检查集群各主机的 SSH 状态

首先查看集群主机是否安装过 SSH（以 YYR 主节点操作为例），同时确认 SSHD 服务的状态是否正常。

检查集群各主机的 SSH 状态

```
# 查看 YYR 主节点的 SSH 安装
[root@YYR ~]# rpm -qa |grep ssh
# 表示已安装了 OpenSSH 客户端
openssh-clients-7.4p1-11.el7.x86_64
libssh2-1.4.3-10.el7_2.1.x86_64
# 表示已安装了 OpenSSH 客户端核心组件
openssh-7.4p1-11.el7.x86_64
# 表示已安装了 OpenSSH 服务端
openssh-server-7.4p1-11.el7.x86_64

# 查看 SSHD 服务是否正常启动
[root@YYR ~]# systemctl status sshd
sshd.service - OpenSSH server daemon
    Loaded:loaded (/usr/lib/systemd/system/sshd.service;enabled;vendor preset:enabled)
    # 表示 SSHD 服务当前为运行状态
    Active:active (running) since — 2020-01-14 08:55:39 CST;52min ago
      Docs:man:sshd(8)
           man:sshd_config(5)
 Main PID:977 (sshd)
    CGroup:/system.slice/sshd.service
           └─977 /usr/sbin/sshd -D

1 月 14 08:55:38YYR systemd[1]:Starting OpenSSH server daemon...
1 月 14 08:55:39 YYR sshd[977]:Server listening on 0.0.0.0 port 22.
1 月 14 08:55:39 YYR sshd[977]:Server listening on::port 22.
1 月 14 08:55:39 YYR systemd[1]:Started OpenSSH server daemon.
1 月 14 09:25:05 YYR sshd[1298]:Accepted password for root from 192.168.3.3 port 60022 ssh2
```

```
1月 14 09:25:05YYR sshd[1300]:Accepted password for root from 192.168.3.3 port 60038 ssh2

# 查看 SSHD 服务是否配置为开机启动
[root@YYR ~]# systemctl list-unit-files |grep sshd
sshd-keygen.service                                   static
# 表示 SSHD 服务被设置为开机启动
sshd.service                                          enabled
sshd@.service                                         static
sshd.socket                                           disabled
[root@YYR ~]#
```

通过以上信息，可以确定以下内容：

■ 运行平台已安装 OpenSSH 的全部组件。

■ SSHD 服务处于运行状态。

■ SSHD 服务被配置为开机启动项。

确认信息后，开始配置免密登录。免密登录配置需要在所有的节点执行操作。

节点免密登录配置

（2）YYR 节点免密登录配置

首先在 YYR 主节点进行配置。

```
# 生成 SSH 密钥
[root@YYR ~]# ssh-keygen -t rsa
Generating public/private rsa key pair.
# 输入密钥存储路径,此处存储在/root/.ssh/id_rsa 目录下
Enter file in which to save the key (/root/.ssh/id_rsa):/root/.ssh/id_rsa
Created directory '/root/.ssh'.
# 直接按 Enter 键
Enter passphrase (empty for no passphrase):
# 再次直接按 Enter 键
Enter same passphrase again:
Your identification has been saved in/root/.ssh/id_rsa.
Your public key has been saved in/root/.ssh/id_rsa.pub.
The key fingerprint is:
SHA256:mv1i246k6H5nZaveDCRyOSsfe+8JRmEaMRpiHaskCXo root@YYR
The key's randomart image is:
```

```
+---[RSA 2048]----+
|. o.o.o          |
|oo .. + o        |
|+ E o . o        |
| + .  =.         |
|   . . * S       |
|    o X  o       |
|   . = *o .      |
|    +. **O..     |
|  o +. + * =BX   |
+----[SHA256]-----+

# 将 SSH 密钥复制到 Superintendent1
[root@YYR ~]# ssh-copy-id Superintendent1
/usr/bin/ssh-copy-id: INFO: Source of key(s) to be installed:"/root/.ssh/id_rsa.pub"
The authenticity of host 'Superintendent1 (192.168.3.191)' can't be established.
ECDSA key fingerprint is SHA256:vI3gIw82+d92BeV1dJ22PbouWYxKqNFX8BwCGxaAvsM.
ECDSA key fingerprint is MD5:8e:4f:7a:11:a5:42:58:08:6b:74:ed:d8:69:78:d1:e3.
# 输入 yes
Are you sure you want to continue connecting (yes/no)? yes
/usr/bin/ssh-copy-id:INFO:attempting to log in with the new key(s),to filter out any that are already installed
/usr/bin/ssh-copy-id:INFO:1 key(s) remain to be installed -- if you are prompted now it is to install the new keys
# 输入 Superintendent1 的 root 密码
root@Superintendent1's password:

Number of key(s) added:1

Now try logging into the machine,with:"ssh 'Superintendent1'"
and check to make sure that only the key(s) you wanted were added.

```

```
# 将 SSH 密钥复制到 Superintendent2
[root@YYR ~]# ssh-copy-id Superintendent2
/usr/bin/ssh-copy-id: INFO: Source of key(s) to be installed:"/root/.ssh/id_rsa.pub"
The authenticity of host 'Superintendent2 (192.168.3.192)' can't be established.
ECDSA key fingerprint is SHA256:QmHhl38QAW+OX1Y5sBrmecuQaaHucokSEvi277VB94w.
ECDSA key fingerprint is MD5:67:31:66:8e:82:7d:67:2c:d5:7d:c7:4e:75:ea:3e:d2.
# 输入 yes
Are you sure you want to continue connecting (yes/no)? yes
/usr/bin/ssh-copy-id:INFO:attempting to log in with the new key(s),to filter out any that are already installed
/usr/bin/ssh-copy-id:INFO:1 key(s) remain to be installed -- if you are prompted now it is to install the new keys
# 输入 Superintendent1 的 root 密码
root@Superintendent2's password:
Number of key(s) added:1
Now try logging into the machine,with:"ssh 'Superintendent2'"
and check to make sure that only the key(s) you wanted were added.
[root@YYR ~]#
```

将公钥添加到授权 Key 列表中。

```
[root@YYR ~]# ls /root/.ssh/
authorized_keys   id_rsa   id_rsa.pub   known_hosts
[root@YYR ~]# cat /root/.ssh/id_rsa.pub >> /root/.ssh/authorized_keys
[root@YYR ~]#
```

下面对 YYR 节点免密登录配置进行验证。

```
# 登录 Superintendent1
[root@YYR ~]# ssh root@Superintendent1
Last login:Mon Jan 13 09:24:43 2020 from 192.168.3.3
```

```
# 显示该信息前不需要输入 Superintendent1 的 root 密码,表示免密登录配置成功
[root@Superintendent1 ~]#
# 退出
[root@Superintendent1 ~]# exit
退出
Connection to Superintendent1 closed.

# 登录 Superintendent2
[root@YYR ~]# ssh root@Superintendent2
Last login:Mon Jan 13 09:25:06 2020 from 192.168.3.3
# 显示该信息前不需要输入 Superintendent1 的 root 密码,表示免密登录配置成功
[root@Superintendent2 ~]#
# 退出
[root@Superintendent2 ~]# exit
退出
Connection to Superintendent2 closed.
[root@YYR ~]#
```

(3) Superintendent1 节点免密登录配置

YYR 节点的免密登录配置成功，接下来配置 Superintendent1。

```
# 生成 SSH 密钥
[root@Superintendent1 ~]# ssh-keygen -t rsa
Generating public/private rsa key pair.
# 输入密钥存储路径,此处存储在/root/.ssh/id_rsa 目录下
Enter file in which to save the key (/root/.ssh/id_rsa):/root/.ssh/id_rsa
# 直接按 Enter 键
Enter passphrase (empty for no passphrase):
# 再次直接按 Enter 键
Enter same passphrase again:
Your identification has been saved in/root/.ssh/id_rsa.
Your public key has been saved in/root/.ssh/id_rsa.pub.
The key fingerprint is:
SHA256:eOKcyMcQNZTa7F7SBdRliXO5Ghml/SirnazjqmRQw5k root@Superintendent1
```

```
The key's randomart image is:
+---[RSA 2048]----+
|     .+o.. + +o  |
|   . +... = + +  |
|    E +   .. =.. |
|  ..oo.   + .o   |
|  . ..o.S..o. .  |
|   o *o+o .o     |
|    =. *o  .     |
|   o . . + .     |
|     ...o+o+     |
+----[SHA256]-----+
# 将 SSH 密钥复制到 YYR
[root@Superintendent2 ~]# ssh-copy-id YYR
/usr/bin/ssh-copy-id:INFO:Source of key(s) to be installed:"/root/.ssh/id_rsa.pub"
The authenticity of host 'YYR (192.168.3.190)' can't be established.
ECDSA key fingerprint is SHA256:kRhfH0fSnLfPXr70iC2QIiPaQJ2S8Hi3NoMRlB0Jm80.
ECDSA key fingerprint is MD5:ba:ec:f6:61:68:ac:c9:a2:f3:06:c5:6c:15:b1:e5:7f.
# 输入 yes
Are you sure you want to continue connecting (yes/no)? yes
/usr/bin/ssh-copy-id:INFO:attempting to log in with the new key(s),to filter out any that are already installed
/usr/bin/ssh-copy-id:INFO:1 key(s) remain to be installed -- if you are prompted now it is to install the new keys
# 输入 YYR 的 root 密码
root@YYR's password:

Number of key(s) added:1

Now try logging into the machine,with:"ssh 'YYR'"
and check to make sure that only the key(s) you wanted were added.
```

```
# 将 SSH 密钥复制到 Superintendent2
[root@Superintendent1 ~]# ssh-copy-id Superintendent2
/usr/bin/ssh-copy-id: INFO: Source of key(s) to be installed:"/root/.ssh/id_rsa.pub"
The authenticity of host 'Superintendent2 (192.168.3.192)' can't be established.
ECDSA key fingerprint is SHA256:QmHhl38QAW+OX1Y5sBrmecuQaaHucokSEvi277VB94w.
ECDSA key fingerprint is MD5:67:31:66:8e:82:7d:67:2c:d5:7d:c7:4e:75:ea:3e:d2.
# 输入 yes
Are you sure you want to continue connecting (yes/no)? yes
/usr/bin/ssh-copy-id:INFO:attempting to log in with the new key(s),to filter out any that are already installed
/usr/bin/ssh-copy-id:INFO:1 key(s) remain to be installed -- if you are prompted now it is to install the new keys
# 输入 Superintendent1 的 root 密码
root@Superintendent2's password:

Number of key(s) added:1

Now try logging into the machine,with:"ssh 'Superintendent2'"
and check to make sure that only the key(s) you wanted were added.
[root@Superintendent1 ~]#
```

将公钥添加到授权 Key 列表中。

```
[root@Superintendent1 ~]# ls /root/.ssh/
authorized_keys   id_rsa   id_rsa.pub   known_hosts
[root@Superintendent1 ~]# cat /root/.ssh/id_rsa.pub >> /root/.ssh/authorized_keys
[root@Superintendent1 ~]#
```

与 YYR 节点相同，对 Superintendent1 节点免密登录配置进行验证。

```
# 登录 YYR
[root@Superintendent1 ~]# ssh root@YYR
Last login:Mon Jan 13 10:13:12 2020 from 192.168.3.191
# 显示该信息前不需要输入 YYR 的 root 密码,表示免密登录配置成功
```

```
[root@YYR ~]#
# 退出
[root@YYR ~]# exit
退出
Connection to YYR closed.
# 登录 Superintendent2
[root@Superintendent1 ~]# ssh root@Superintendent2
Last login:Mon Jan 13 10:08:44 2020 from 192.168.3.190
# 显示该信息前不需要输入 Superintendent2 的 root 密码,表示免密登录配置成功
[root@Superintendent2 ~]#
# 退出
[root@Superintendent2 ~]# exit
退出
Connection to Superintendent2 closed.
[root@Superintendent1 ~]#
```

（4）Superintendent2 节点免密登录配置

Superintendent1 节点的免密登录配置成功后，通过相同的操作来配置 Superintendent2。

```
# 生成 SSH 密钥
[root@Superintendent2 ~]# ssh-keygen -t rsa
Generating public/private rsa key pair.
# 输入密钥存储路径,此处存储在/root/.ssh/id_rsa 目录下
Enter file in which to save the key (/root/.ssh/id_rsa):/root/.ssh/id_rsa
# 直接按 Enter 键
Enter passphrase (empty for no passphrase):
# 再次直接按 Enter 键
Enter same passphrase again:
Your identification has been saved in/root/.ssh/id_rsa.
Your public key has been saved in/root/.ssh/id_rsa.pub.
The key fingerprint is:
SHA256:AaAeXcg5XpUpGvGuzJXUppwVpRm+KtQCOB/P6Juubpw root@Superintendent2
The key's randomart image is:
+---[RSA 2048]----+
```

```
|   .o=o. ++.       |
|. o*oo+o=          |
|o=.. =o.O          |
| + Bo= * o         |
|   + =O S          |
|. + + .            |
|. o=.              |
|E o .              |
|+ + +              |
+----[SHA256]-----+
# 将 SSH 密钥复制到 YYR
[root@Superintendent2 ~]# ssh-copy-id YYR
/usr/bin/ssh-copy-id:INFO:Source of key(s) to be installed:"/root/.ssh/id_rsa.pub"
The authenticity of host 'YYR (192.168.3.190)' can't be established.
ECDSA key fingerprint is SHA256:kRhfH0fSnLfPXr70iC2QIiPaQJ2S8Hi3NoMR1B0Jm80.
ECDSA key fingerprint is MD5:ba:ec:f6:61:68:ac:c9:a2:f3:06:c5:6c:15:b1:e5:7f.
# 输入 yes
Are you sure you want to continue connecting (yes/no)? yes
/usr/bin/ssh-copy-id:INFO:attempting to log in with the new key(s),to filter out any that are already installed
/usr/bin/ssh-copy-id:INFO:1 key(s) remain to be installed -- if you are prompted now it is to install the new keys
# 输入 YYR 的 root 密码
root@YYR's password:
Number of key(s) added:1
Now try logging into the machine,with:"ssh 'YYR'"
and check to make sure that only the key(s) you wanted were added.
# 将 SSH 密钥复制到 Superintendent1
[root@Superintendent2 ~]# ssh-copy-id Superintendent1
```

```
/usr/bin/ssh-copy-id:INFO:Source of key(s) to be installed:"/root/.ssh/id_rsa.pub"
The authenticity of host 'Superintendent1 (192.168.3.191)' can't be established.
ECDSA key fingerprint is SHA256:vI3gIw82+d92BeV1dJ22PbouWYxKqNFX8BwCGxaAvsM.
ECDSA key fingerprint is MD5:8e:4f:7a:11:a5:42:58:08:6b:74:ed:d8:69:78:d1:e3.
# 输入 yes
Are you sure you want to continue connecting (yes/no)? yes
/usr/bin/ssh-copy-id:INFO:attempting to log in with the new key(s),to filter out any that are already installed
/usr/bin/ssh-copy-id:INFO:1 key(s) remain to be installed -- if you are prompted now it is to install the new keys
# 输入 Superintendent1 的 root 密码
root@Superintendent1's password:
Number of key(s) added:1
Now try logging into the machine,with:"ssh 'Superintendent1'"
and check to make sure that only the key(s) you wanted were added.
[root@Superintendent2 ~]#
```

将公钥添加到授权 Key 列表中。

```
[root@Superintendent2 ~]# ls /root/.ssh/
authorized_keys  id_rsa  id_rsa.pub  known_hosts
[root@Superintendent2 ~]# cat /root/.ssh/id_rsa.pub >> /root/.ssh/authorized_keys
[root@Superintendent2 ~]#
```

对 Superintendent2 节点免密登录配置进行验证。

```
# 登录 YYR
[root@Superintendent2 ~]# ssh root@YYR
Last login:Mon Jan 13 10:46:08 2020 from 192.168.3.191
# 显示该信息前不需要输入 YYR 的 root 密码,表示免密登录配置成功
[root@YYR ~]#
```

```
# 退出
[root@YYR ~]# exit
退出
Connection to YYR closed.
# 登录 Superintendent1
[root@Superintendent2 ~]# ssh root@Superintendent1
Last login:Mon Jan 13 10:07:57 2020 from 192.168.3.190
# 显示该信息前不需要输入 Superintendent1 的 root 密码,表示免密登录配置成功
[root@Superintendent1 ~]#
# 退出
[root@Superintendent1 ~]# exit
退出
Connection to Superintendent1 closed.
[root@Superintendent2 ~]#
```

至此，集群内全部主机节点之间的免密登录配置完成。下面开始对 JDK 进行安装。

2. 安装 JDK

根据 Hadoop 2.6.0 的官方开发社区描述，选择 JDK8 64 位版本作为 Hadoop 的基础环境，需要下载 Linux X64 位版本的 jdk-8u191-linux-x64.tar.gz 文件。下载链接如下：

https://www.oracle.com/technetwork/java/javase/downloads/jdk8-downloads-2133151.html

（1）查看集群主机是否安装过 JDK

很多操作系统默认会安装 OpenJDK，但由于 OpenJDK 采用 GPL 协议，它的源代码不完整。尽管 Hadoop 官方声明其对 OpenJDK 提供了支持，但是官方还是建议使用 Oracle 官方发布的稳定版 JDK。

CentOS 7 Mini 版操作系统默认并不会安装 OpenJDK，为了确保运行平台环境的正确性，在安装 JDK 之前，还要对 JDK 环境进行确认。

```
[root@YYR ~]# rpm -qa |grep jdk
# YYR 节点没有安装 JDK
[root@YYR ~]#

[root@Superintendent1 ~]# rpm -qa |grep jdk
# Superintendent1 节点没有安装 JDK
[root@Superintendent1 ~]#

```

安装 JDK 前的准备工作

```
[root@Superintendent2 ~]# rpm -qa |grep jdk
# Superintendent2 节点没有安装 JDK
[root@Superintendent2 ~]#
```

（2）安装 JDK

首先将下载好的 JDK 安装包上传到 YYR 节点的 TMP 路径下（因为已经在集群内所有的节点配置了 SSH，可以使用 SSH、SFTP 等多种方式上传，此处推荐使用开源远程访问软件 MobaXterm 进行文件上传）。

之后将 jdk-8u191-linux-x64. tar. gz 文件解压到/usr/share/java 路径下。

```
# 创建文件夹/usr/share/java
[root@YYR ~]# mkdir /usr/share/java
# 将 JDK 解压到/usr/share/java 文件夹下
[root@YYR ~]# tar -zxvf /tmp/jdk-8u191-linux-x64.tar.gz -C /usr/share/java

# 查看/usr/share/java 文件夹下的目录结构
[root@YYR ~]# ls /usr/share/java/
jdk1.8.0_191
```

解压完成后，对 YYR 的 JDK 环境进行配置。

```
# 检查确认 JDK 为未生效状态
[root@YYR ~]# java -version
-bash:java:未找到命令

# 编辑配置文件
[root@YYR ~]# sudo vi /etc/profile

# 在配置文件最下方增加以下内容:
#set java environment
# JDK 的解压路径
export JAVA_HOME=/usr/share/java/jdk1.8.0_191
export JRE_HOME=${JAVA_HOME}/jre
export CLASSPATH=.:${JAVA_HOME}/lib:${JRE_HOME}/lib
export PATH=$PATH:${JAVA_HOME}/bin:${JRE_HOME}/bin

# 执行 profile 文件
[root@YYR ~]# source /etc/profile
```

```
# 检查新安装的 JDK 是否生效
[root@YYR ~]# java -version
java version "1.8.0_191"
Java(TM) SE Runtime Environment (build 1.8.0_191-b12)
Java HotSpot(TM) 64-Bit Server VM (build 25.191-b12, mixed mode)
[root@YYR ~]#
```

至此，YYR 节点的 JDK 环境部署完成。使用 SSH 的功能快速完成 Superintendent1、Superintendent2 的 JDK 安装部署工作。Superintendent1、Superintendent2 两台节点的安装部署工作只需要操作 YYR 主机即可完成。

```
# 备份 Superintendent1 的/etc/profile 文件
[root@YYR ~]# ssh root@Superintendent1
Last login:Mon Jan 13 11:00:41 2020 from 192.168.3.192
[root@Superintendent1 ~]# cp/etc/profile/etc/profile.bak
[root@Superintendent1 ~]# exit
退出
Connection to Superintendent1 closed.
# 备份 Superintendent2 的/etc/profile 文件
[root@YYR ~]# ssh root@Superintendent2
Last login:Mon Jan 13 10:46:16 2020 from 192.168.3.191
[root@Superintendent2 ~]# cp/etc/profile/etc/profile.bak
[root@Superintendent2 ~]# exit
退出
Connection to Superintendent2 closed.
# 将解压后的 Java 文件夹分别拷贝到 Superintendent1、Superintendent2
[root@YYR ~]# scp -r/usr/share/java root@Superintendent1:/usr/share/java
[root@YYR ~]# scp -r/usr/share/java root@Superintendent2:/usr/share/java
# 将解 YYR 配置好的/etc/profile 文件分别拷贝到 Superintendent1、Superintendent2
[root@YYR ~]# scp/etc/profile root@Superintendent1:/etc/profile
```

```
profile                                    100% 2021   116.3KB/s   00:00
[root@YYR ~]# scp/etc/profile root@Superintendent2:/etc/profile
profile                                    100% 2021   107.8KB/s   00:00

# 执行 Superintendent1 的/etc/profile 并验证 JDK 安装
[root@YYR ~]# ssh root@Superintendent1
Last login:Mon Jan 13 13:00:25 2020 from 192.168.3.190
[root@Superintendent1 ~]# source/etc/profile
[root@Superintendent1 ~]# java -version
java version "1.8.0_191"
Java(TM) SE Runtime Environment (build 1.8.0_191 -b12)
Java HotSpot(TM) 64 - Bit Server VM (build 25.191 - b12,mixed mode)
[root@Superintendent1 ~]# exit
退出
Connection to Superintendent1 closed.

# 执行 Superintendent2 的/etc/profile 并验证 JDK 安装
[root@YYR ~]# ssh root@Superintendent2
Last login:Mon Jan 13 13:00:53 2020 from 192.168.3.190
[root@Superintendent2 ~]# source/etc/profile
[root@Superintendent2 ~]# java -version
java version "1.8.0_191"
Java(TM) SE Runtime Environment (build 1.8.0_191 -b12)
Java HotSpot(TM) 64 - Bit Server VM (build 25.191 - b12,mixed mode)
[root@Superintendent2 ~]# exit
退出
Connection to Superintendent2 closed.
[root@YYR ~]#
```

现在已经完成了 Hadoop 安装前的全部准备工作，下面来完成 Hadoop 的完全分布式安装部署。

任务5　安装配置 Hadoop

在一系列准备工作完成后，即将开始 Hadoop 完全分布式的搭建工作。请通过以下官方

URL 下载完整的 Hadoop 2.6.0 软件包：

https://archive.apache.org/dist/hadoop/common/hadoop-2.6.0/hadoop-2.6.0.tar.gz

1. 安装 Hadoop

安装 HADOOP

将下载的 Hadoop 软件包上传到 YYR 节点的 tmp 路径下，解压缩后，使用 SSH 命令将 Hadoop 拷贝到各 Slave 节点的相同位置。

```
# 将 Hadoop 软件包解到/usr/local/
[root@YYR ~]# tar -zxvf/tmp/hadoop-2.6.0.tar.gz -C/usr/local/

# 解压后的 Hadoop 文件夹为 hadoop-2.6.0
[root@YYR ~]]# ls/usr/local/
bin  etc  games  hadoop-2.6.0  include  lib  lib64  libexec  sbin  share  src

# 将 Hadoop 文件夹名称修改为 hadoop
[root@YYR ~]]# mv/usr/local/hadoop-2.6.0 /usr/local/hadoop

# 将 Hadoop 文件夹拷贝到 Superintendent1 相同位置
[root@YYR ~]# scp -r/usr/local/hadoop/root@Superintendent1:/usr/local/

# 将 Hadoop 文件夹拷贝到 Superintendent2 相同位置
[root@YYR ~]# scp -r/usr/local/hadoop/root@Superintendent2:/usr/local/
```

2. 环境变量配置

```
# 修改/etc/profile
[root@YYR ~]# sudo vi/etc/profile

export JAVA_HOME=/usr/share/java/jdk1.8.0_191
export JRE_HOME=${JAVA_HOME}/jre
export CLASSPATH=.:${JAVA_HOME}/lib:${JRE_HOME}/lib
# 新添加内容
export HADOOP_HOME=/usr/local/hadoop
```

```
# 新增内容: ${HADOOP_HOME}/bin: ${HADOOP_HOME}/sbin
export PATH=$PATH:${JAVA_HOME}/bin:${JRE_HOME}/bin:${HADOOP_HOME}/bin:${HADOOP_HOME}/sbin

# 执行/etc/profile
[root@YYR ~]# source /etc/profile
# 验证配置文件正确
[root@YYR ~]# hadoop version
Hadoop 2.6.0
Subversion https://git-wip-us.apache.org/repos/asf/hadoop.git -r e3496499ecb8d220fba99dc5ed4c99c8f9e33bb1
Compiled by jenkins on 2014-11-13T21:10Z
Compiled with protoc 2.5.0
From source with checksum 18e43357c8f927c0695f1e9522859d6a
This command was run using /usr/local/hadoop/share/hadoop/common/hadoop-common-2.6.0.jar

# 将/etc/profile 发送到 Salve1 相同位置
[root@YYR ~]# scp /etc/profile root@Superintendent1:/etc/profile
profile                                   100% 2097   128.2KB/s   00:00

# 将/etc/profile 发送到 Salve2 相同位置
[root@YYR ~]# scp /etc/profile root@Superintendent2:/etc/profile
profile                                   100% 2097   152.2KB/s   00:00

# 登录 Superintendent1,执行/etc/profile,验证 Hadoop 配置
[root@YYR ~]# ssh root@Superintendent1
Last login:Mon Jan 13 14:12:37 2020 from 192.168.3.3
[root@Superintendent1 ~]# source /etc/profile
[root@Superintendent1 ~]# hadoop version
Hadoop 2.6.0
Subversion https://git-wip-us.apache.org/repos/asf/hadoop.git -r e3496499ecb8d220fba99dc5ed4c99c8f9e33bb1
Compiled by jenkins on 2014-11-13T21:10Z
Compiled with protoc 2.5.0
```

```
From source with checksum 18e43357c8f927c0695f1e9522859d6a
This command was run using /usr/local/hadoop/share/hadoop/common/hadoop-common-2.6.0.jar
[root@Superintendent1 ~]# exit
退出
Connection to Superintendent1 closed.

# 登录 Superintendent2,执行/etc/profile,验证 Hadoop 配置
[root@YYR ~]# ssh root@Superintendent2
Last login:Mon Jan 13 14:12:51 2020 from 192.168.3.3
[root@Superintendent2 ~]# source /etc/profile
[root@Superintendent2 ~]# hadoop version
Hadoop 2.6.0
Subversion https://git-wip-us.apache.org/repos/asf/hadoop.git -r e3496499ecb8d220fba99dc5ed4c99c8f9e33bb1
Compiled by jenkins on 2014-11-13T21:10Z
Compiled with protoc 2.5.0
From source with checksum 18e43357c8f927c0695f1e9522859d6a
This command was run using /usr/local/hadoop/share/hadoop/common/hadoop-common-2.6.0.jar
[root@Superintendent2 ~]# exit
退出
Connection to Superintendent2 closed.
[root@YYR ~]#
```

在集群环境下，即使各结点都正确地配置了 JAVA_HOME，Hadoop 启动时也会报错，需要在 Hadoop-env. sh 中也指定 Java 路径。

```
[root@YYR ~]# sudo vi /usr/local/hadoop/etc/hadoop/hadoop-env.sh

# 将配置文件中的 JAVA_HOME 修改为如下配置
export JAVA_HOME=/usr/share/java/jdk1.8.0_191

# 发送到各 Slave 节点
[root@YYR ~]# scp /usr/local/hadoop/etc/hadoop/hadoop-env.sh root@Superintendent1:/usr/local/hadoop/etc/hadoop/hadoop-env.sh
```

```
hadoop-env.sh                   100% 4240   228.0KB/s   00:00

[root@YYR ~]# scp /usr/local/hadoop/etc/hadoop/hadoop-env.sh root@Superintendent2:/usr/local/hadoop/etc/hadoop/hadoop-env.sh
hadoop-env.sh                   100% 4240     5.1MB/s   00:00
```

3. 配置主节点 Hadoop 运行参数

首先，在主节点创建一个 tmp 目录，创建方式如下：

```
[root@YYR ~]# mkdir /usr/local/hadoop/tmp
[root@YYR ~]# chmod 777 /usr/local/hadoop/tmp
```

Hadoop 的运行参数配置文件存放在/usr/local/hadoop/etc/hadoop 目录下，为了能够使 Hadoop 正常运行，需要配置以下五个文件的参数：

- slaves
- core-site. xml
- hdfs-site. xml
- mapred-site. xml（注意，该目录下是没有这个文件的，需要根据 mapred-site. xml. template 创建）
- yarn-site. xml

（1）配置 slaves 文件

```
[root@YYR ~]# sudo vi /usr/local/hadoop/etc/hadoop/slaves

# 修改为如下配置,此处配置的节点是所有 DataNode 主机名
YYR
Superintendent1
Superintendent2
```

配置 SLAVES 文件

（2）配置 core-site. xml 文件

```
[root@YYR ~]# sudo vi /usr/local/hadoop/etc/hadoop/core-site.xml

<?xml version="1.0" encoding="UTF-8"?>
<?xml-stylesheet type="text/xsl" href="configuration.xsl"?>
<!--
  Licensed under the Apache License,Version 2.0 (the "License");
```

```
  you may not use this file except in compliance with the License.
  You may obtain a copy of the License at

    http://www.apache.org/licenses/LICENSE-2.0

  Unless required by applicable law or agreed to in writing, software
  distributed under the License is distributed on an "AS IS" BASIS,
  WITHOUT WARRANTIES OR CONDITIONS OF ANY KIND, either express or implied.
  See the License for the specific language governing permissions and
  limitations under the License. See accompanying LICENSE file.
-->

<!-- Put site-specific property overrides in this file. -->

<configuration>
<!-->以下为新添加项<-->
    <property>
        <name>fs.default.name</name>
        <value>hdfs://YYR:9000</value>
    </property>
    <property>
        <name>hadoop.tmp.dir</name>
        <value>/usr/local/hadoop/tmp</value>
    </property>
    <property>
        <name>fs.trash.interval</name>
        <value>10080</value>
    </property>
<!-->以上添加内容结束于此<-->
</configuration>
```

(3) 配置 hdfs-site.xml 文件

```
[root@YYR ~]# sudo vi /usr/local/hadoop/etc/hadoop/hdfs-site.xml
<?xml version="1.0" encoding="UTF-8"?>
<?xml-stylesheet type="text/xsl" href="configuration.xsl"?>
<!--
  Licensed under the Apache License,Version 2.0 (the "License");
  you may not use this file except in compliance with the License.
  You may obtain a copy of the License at
    http://www.apache.org/licenses/LICENSE-2.0
  Unless required by applicable law or agreed to in writing,software
  distributed under the License is distributed on an "AS IS" BASIS,
  WITHOUT WARRANTIES OR CONDITIONS OF ANY KIND,either express or implied.
  See the License for the specific language governing permissions and
  limitations under the License. See accompanying LICENSE file.
-->
<!-- Put site-specific property overrides in this file. -->
<configuration>
<!-->以下为新添加项<-->
    <property>
        <name>dfs.namenode.secondary.http-address</name>
        <value>YYR:50090</value>
    </property>
    <property>
        <name>dfs.replication</name>
        <!-->副本个数<-->
        <value>3</value>
```

```
    </property>
    <property>
        <name>dfs.namenode.name.dir</name>
        <value>/usr/local/hadoop/hdfs/name</value>
    </property>
    <property>
        <name>dfs.datanode.data.dir</name>
        <value>/usr/local/hadoop/hdfs/data</value>
    </property>
<!--> 以上添加内容结束于此 <-->
</configuration>
```

（4）配置 mapred-site. xml 文件

```
# 使用官方文件模板创建配置文件
[root@YYR ~]# cp /usr/local/hadoop/etc/hadoop/mapred-site.xml.template /usr/local/hadoop/etc/hadoop/mapred-site.xml

# 修改配置文件内容
[root@YYR ~]# sudo vi /usr/local/hadoop/etc/hadoop/mapred-site.xml

<?xml version="1.0"?>
<?xml-stylesheet type="text/xsl" href="configuration.xsl"?>
<!--
  Licensed under the Apache License, Version 2.0 (the "License");
  you may not use this file except in compliance with the License.
  You may obtain a copy of the License at

    http://www.apache.org/licenses/LICENSE-2.0

  Unless required by applicable law or agreed to in writing, software
  distributed under the License is distributed on an "AS IS" BASIS,
```

```
WITHOUT WARRANTIES OR CONDITIONS OF ANY KIND,either express or implied.
 See the License for the specific language governing permissions and
limitations under the License. See accompanying LICENSE file.
-->

<!-- Put site-specific property overrides in this file. -->

<configuration>
<!-->以下为新添加项<-->
    <property>
        <name>mapreduce.framework.name</name>
        <value>yarn</value>
    </property>
<!-->以上添加内容结束于此<-->
</configuration>
```

（5）配置 yarn-site. xml 文件

```
[root@YYR ~]# sudo vi /usr/local/hadoop/etc/hadoop/yarn-site.xml

<?xml version="1.0"?>
<!--
  Licensed under the Apache License,Version 2.0 (the "License");
  you may not use this file except in compliance with the License.
  You may obtain a copy of the License at

    http://www.apache.org/licenses/LICENSE-2.0

  Unless required by applicable law or agreed to in writing,software
  distributed under the License is distributed on an "AS IS" BASIS,
  WITHOUT WARRANTIES OR CONDITIONS OF ANY KIND,either express or implied.
```

```
  See the License for the specific language governing permissions and
 limitations under the License. See accompanying LICENSE file.
-->
<configuration>

<!--Site specific YARN configuration properties -->
<!-->以下为新添加项<-->
    <property>
        <name>yarn.nodemanager.aux-services</name>
        <value>mapreduce_shuffle</value>
    </property>
<!-->以上添加内容结束于此<-->
</configuration>
```

（6）初始化 NameNode

```
# 运行 NameNode 格式化命令
[root@YYR ~]# hadoop namenode -format
DEPRECATED:Use of this script to execute hdfs command is deprecated.
Instead use the hdfs command for it.

20/01/13 15:08:05 INFO namenode.NameNode:STARTUP_MSG:
/************************************************************
STARTUP_MSG:Starting NameNode
STARTUP_MSG:host =YYR/192.168.3.190
STARTUP_MSG:args =[-format]
STARTUP_MSG:version =2.6.0
STARTUP_MSG:classpath =/usr/local/hadoop/etc/hadoop:/usr/local/hadoop/share/hadoop/common/lib/jackson-xc-1.9.13.jar:/usr/local/hadoop/share/hadoop/common/lib/netty-3.6.2.Final.jar:/usr/local/hadoop/share/hadoop/common/lib/jackson-core-asl-1.9.13.jar:/usr/local/hadoop/share/hadoop/common/lib/snappy-java-1.0.4.1.jar:/usr/local/hadoop/share/hadoop/common/lib/commons-collections-3.2.1.jar:/usr/local/hadoop/share/hadoop/common/lib/jersey-json-
```

1.9.jar:/usr/local/hadoop/share/hadoop/common/lib/commons-
beanutils-1.7.0.jar:/usr/local/hadoop/share/hadoop/com-
mon/lib/slf4j-log4j12-1.7.5.jar:/usr/local/hadoop/share/
hadoop/common/lib/apacheds-i18n-2.0.0-M15.jar:/usr/lo-
cal/hadoop/share/hadoop/common/lib/hadoop-auth-2.6.0.
jar:/usr/local/hadoop/share/hadoop/common/lib/hadoop-an-
notations-2.6.0.jar:/usr/local/hadoop/share/hadoop/com-
mon/lib/commons-codec-1.4.jar:/usr/local/hadoop/share/
hadoop/common/lib/httpcore-4.2.5.jar:/usr/local/hadoop/
share/hadoop/common/lib/commons-digester-1.8.jar:/usr/lo-
cal/hadoop/share/hadoop/common/lib/asm-3.2.jar:/usr/lo-
cal/hadoop/share/hadoop/common/lib/jersey-core-1.9.jar:/
usr/local/hadoop/share/hadoop/common/lib/jettison-
1.1.jar:/usr/local/hadoop/share/hadoop/common/lib/jersey-
server-1.9.jar:/usr/local/hadoop/share/hadoop/common/lib/
jetty-util-6.1.26.jar:/usr/local/hadoop/share/hadoop/com-
mon/lib/java-xmlbuilder-0.4.jar:/usr/local/hadoop/share/
hadoop/common/lib/jaxb-api-2.2.2.jar:/usr/local/hadoop/
share/hadoop/common/lib/curator-recipes-2.6.0.jar:/usr/
local/hadoop/share/hadoop/common/lib/api-asn1-api-1.0.0-
M20.jar:/usr/local/hadoop/share/hadoop/common/lib/gson-
2.2.4.jar:/usr/local/hadoop/share/hadoop/common/lib/cura-
tor-client-2.6.0.jar:/usr/local/hadoop/share/hadoop/com-
mon/lib/commons-lang-2.6.jar:/usr/local/hadoop/share/ha-
doop/common/lib/curator-framework-2.6.0.jar:/usr/local/
hadoop/share/hadoop/common/lib/slf4j-api-1.7.5.jar:/usr/
local/hadoop/share/hadoop/common/lib/guava-11.0.2.jar:/
usr/local/hadoop/share/hadoop/common/lib/junit-4.11.jar:/
usr/local/hadoop/share/hadoop/common/lib/api-util-1.0.0-
M20.jar:/usr/local/hadoop/share/hadoop/common/lib/jackson-
jaxrs-1.9.13.jar:/usr/local/hadoop/share/hadoop/common/
lib/httpclient-4.2.5.jar:/usr/local/hadoop/share/hadoop/
common/lib/jetty-6.1.26.jar:/usr/local/hadoop/share/ha-
doop/common/lib/jasper-runtime-5.5.23.jar:/usr/local/ha-
doop/share/hadoop/common/lib/commons-httpclient-3.1.jar:/
usr/local/hadoop/share/hadoop/common/lib/stax-api-1.0-
2.jar:/usr/local/hadoop/share/hadoop/common/lib/htrace-

core -3.0.4.jar:/usr/local/hadoop/share/hadoop/common/lib/commons - compress - 1.4.1.jar:/usr/local/hadoop/share/hadoop/common/lib/commons - logging - 1.1.3.jar:/usr/local/hadoop/share/hadoop/common/lib/hamcrest - core - 1.3.jar:/usr/local/hadoop/share/hadoop/common/lib/commons - net - 3.1.jar:/usr/local/hadoop/share/hadoop/common/lib/apacheds - kerberos - codec - 2.0.0 - M15.jar:/usr/local/hadoop/share/hadoop/common/lib/commons - io - 2.4.jar:/usr/local/hadoop/share/hadoop/common/lib/servlet - api - 2.5.jar:/usr/local/hadoop/share/hadoop/common/lib/xmlenc - 0.52.jar:/usr/local/hadoop/share/hadoop/common/lib/commons - cli - 1.2.jar:/usr/local/hadoop/share/hadoop/common/lib/jsr305 - 1.3.9.jar:/usr/local/hadoop/share/hadoop/common/lib/jackson - mapper - asl - 1.9.13.jar:/usr/local/hadoop/share/hadoop/common/lib/commons - math3 - 3.1.1.jar:/usr/local/hadoop/share/hadoop/common/lib/paranamer - 2.3.jar:/usr/local/hadoop/share/hadoop/common/lib/commons - el - 1.0.jar:/usr/local/hadoop/share/hadoop/common/lib/jasper - compiler - 5.5.23.jar:/usr/local/hadoop/share/hadoop/common/lib/jsch - 0.1.42.jar:/usr/local/hadoop/share/hadoop/common/lib/protobuf - java - 2.5.0.jar:/usr/local/hadoop/share/hadoop/common/lib/jaxb - impl - 2.2.3 - 1.jar:/usr/local/hadoop/share/hadoop/common/lib/xz - 1.0.jar:/usr/local/hadoop/share/hadoop/common/lib/jsp - api - 2.1.jar:/usr/local/hadoop/share/hadoop/common/lib/mockito - all - 1.8.5.jar:/usr/local/hadoop/share/hadoop/common/lib/commons - configuration - 1.6.jar:/usr/local/hadoop/share/hadoop/common/lib/zookeeper - 3.4.6.jar:/usr/local/hadoop/share/hadoop/common/lib/avro - 1.7.4.jar:/usr/local/hadoop/share/hadoop/common/lib/commons - beanutils - core - 1.8.0.jar:/usr/local/hadoop/share/hadoop/common/lib/jets3t - 0.9.0.jar:/usr/local/hadoop/share/hadoop/common/lib/log4j - 1.2.17.jar:/usr/local/hadoop/share/hadoop/common/lib/activation - 1.1.jar:/usr/local/hadoop/share/hadoop/common/hadoop - common - 2.6.0 - tests.jar:/usr/local/hadoop/share/hadoop/common/hadoop - common - 2.6.0.jar:/usr/local/hadoop/share/hadoop/common/hadoop - nfs - 2.6.0.jar:/usr/local/hadoop/share/hadoop/

```
hdfs:/usr/local/hadoop/share/hadoop/hdfs/lib/netty-3.6.2.Final.jar:/usr/local/hadoop/share/hadoop/hdfs/lib/jackson-core-asl-1.9.13.jar:/usr/local/hadoop/share/hadoop/hdfs/lib/commons-codec-1.4.jar:/usr/local/hadoop/share/hadoop/hdfs/lib/asm-3.2.jar:/usr/local/hadoop/share/hadoop/hdfs/lib/jersey-core-1.9.jar:/usr/local/hadoop/share/hadoop/hdfs/lib/jersey-server-1.9.jar:/usr/local/hadoop/share/hadoop/hdfs/lib/jetty-util-6.1.26.jar:/usr/local/hadoop/share/hadoop/hdfs/lib/commons-lang-2.6.jar:/usr/local/hadoop/share/hadoop/hdfs/lib/guava-11.0.2.jar:/usr/local/hadoop/share/hadoop/hdfs/lib/xml-apis-1.3.04.jar:/usr/local/hadoop/share/hadoop/hdfs/lib/jetty-6.1.26.jar:/usr/local/hadoop/share/hadoop/hdfs/lib/jasper-runtime-5.5.23.jar:/usr/local/hadoop/share/hadoop/hdfs/lib/htrace-core-3.0.4.jar:/usr/local/hadoop/share/hadoop/hdfs/lib/commons-logging-1.1.3.jar:/usr/local/hadoop/share/hadoop/hdfs/lib/xercesImpl-2.9.1.jar:/usr/local/hadoop/share/hadoop/hdfs/lib/commons-io-2.4.jar:/usr/local/hadoop/share/hadoop/hdfs/lib/servlet-api-2.5.jar:/usr/local/hadoop/share/hadoop/hdfs/lib/xmlenc-0.52.jar:/usr/local/hadoop/share/hadoop/hdfs/lib/commons-cli-1.2.jar:/usr/local/hadoop/share/hadoop/hdfs/lib/jsr305-1.3.9.jar:/usr/local/hadoop/share/hadoop/hdfs/lib/jackson-mapper-asl-1.9.13.jar:/usr/local/hadoop/share/hadoop/hdfs/lib/commons-el-1.0.jar:/usr/local/hadoop/share/hadoop/hdfs/lib/protobuf-java-2.5.0.jar:/usr/local/hadoop/share/hadoop/hdfs/lib/commons-daemon-1.0.13.jar:/usr/local/hadoop/share/hadoop/hdfs/lib/jsp-api-2.1.jar:/usr/local/hadoop/share/hadoop/hdfs/lib/log4j-1.2.17.jar:/usr/local/hadoop/share/hadoop/hdfs/hadoop-hdfs-nfs-2.6.0.jar:/usr/local/hadoop/share/hadoop/hdfs/hadoop-hdfs-2.6.0-tests.jar:/usr/local/hadoop/share/hadoop/hdfs/hadoop-hdfs-2.6.0.jar:/usr/local/hadoop/share/hadoop/yarn/lib/jackson-xc-1.9.13.jar:/usr/local/hadoop/share/hadoop/yarn/lib/netty-3.6.2.Final.jar:/usr/local/hadoop/share/hadoop/yarn/lib/jackson-core-asl-1.9.13.jar:/usr/local/hadoop/share/hadoop/yarn/
```

lib/commons - collections - 3.2.1.jar:/usr/local/hadoop/share/hadoop/yarn/lib/jersey - json -1.9.jar:/usr/local/hadoop/share/hadoop/yarn/lib/commons - codec - 1.4.jar:/usr/local/hadoop/share/hadoop/yarn/lib/jline -0.9.94.jar:/usr/local/hadoop/share/hadoop/yarn/lib/asm -3.2.jar:/usr/local/hadoop/share/hadoop/yarn/lib/jersey - core -1.9.jar:/usr/local/hadoop/share/hadoop/yarn/lib/jettison -1.1.jar:/usr/local/hadoop/share/hadoop/yarn/lib/jersey - server -1.9.jar:/usr/local/hadoop/share/hadoop/yarn/lib/jetty -util -6.1.26.jar:/usr/local/hadoop/share/hadoop/yarn/lib/jaxb - api -2.2.2.jar:/usr/local/hadoop/share/hadoop/yarn/lib/aopalliance -1.0.jar:/usr/local/hadoop/share/hadoop/yarn/lib/commons - lang -2.6.jar:/usr/local/hadoop/share/hadoop/yarn/lib/guava -11.0.2.jar:/usr/local/hadoop/share/hadoop/yarn/lib/jersey - guice -1.9.jar:/usr/local/hadoop/share/hadoop/yarn/lib/javax.inject -1.jar:/usr/local/hadoop/share/hadoop/yarn/lib/jackson - jaxrs -1.9.13.jar:/usr/local/hadoop/share/hadoop/yarn/lib/jetty -6.1.26.jar:/usr/local/hadoop/share/hadoop/yarn/lib/commons - httpclient -3.1.jar:/usr/local/hadoop/share/hadoop/yarn/lib/stax - api -1.0 -2.jar:/usr/local/hadoop/share/hadoop/yarn/lib/commons -compress -1.4.1.jar:/usr/local/hadoop/share/hadoop/yarn/lib/leveldbjni - all -1.8.jar:/usr/local/hadoop/share/hadoop/yarn/lib/commons - logging -1.1.3.jar:/usr/local/hadoop/share/hadoop/yarn/lib/guice -3.0.jar:/usr/local/hadoop/share/hadoop/yarn/lib/jersey - client -1.9.jar:/usr/local/hadoop/share/hadoop/yarn/lib/commons - io -2.4.jar:/usr/local/hadoop/share/hadoop/yarn/lib/servlet - api -2.5.jar:/usr/local/hadoop/share/hadoop/yarn/lib/commons -cli - 1.2.jar:/usr/local/hadoop/share/hadoop/yarn/lib/jsr305 -1.3.9.jar:/usr/local/hadoop/share/hadoop/yarn/lib/jackson - mapper - asl -1.9.13.jar:/usr/local/hadoop/share/hadoop/yarn/lib/protobuf - java -2.5.0.jar:/usr/local/hadoop/share/hadoop/yarn/lib/jaxb - impl -2.2.3 -1.jar:/usr/local/hadoop/share/hadoop/yarn/lib/xz -1.0.jar:/usr/local/hadoop/share/hadoop/yarn/lib/zookeeper -3.4.6.jar:/usr/local/hadoop/share/hadoop/yarn/lib/log4j -1.2.17.jar:/usr/

local/hadoop/share/hadoop/yarn/lib/guice-servlet-
3.0.jar:/usr/local/hadoop/share/hadoop/yarn/lib/activa-
tion-1.1.jar:/usr/local/hadoop/share/hadoop/yarn/hadoop-
yarn-api-2.6.0.jar:/usr/local/hadoop/share/hadoop/yarn/
hadoop-yarn-server-resourcemanager-2.6.0.jar:/usr/local/
hadoop/share/hadoop/yarn/hadoop-yarn-client-2.6.0.jar:/
usr/local/hadoop/share/hadoop/yarn/hadoop-yarn-server-
applicationhistoryservice-2.6.0.jar:/usr/local/hadoop/
share/hadoop/yarn/hadoop-yarn-server-web-proxy-2.6.0.
jar:/usr/local/hadoop/share/hadoop/yarn/hadoop-yarn-
server-common-2.6.0.jar:/usr/local/hadoop/share/hadoop/
yarn/hadoop-yarn-server-nodemanager-2.6.0.jar:/usr/lo-
cal/hadoop/share/hadoop/yarn/hadoop-yarn-registry-
2.6.0.jar:/usr/local/hadoop/share/hadoop/yarn/hadoop-yarn-
server-tests-2.6.0.jar:/usr/local/hadoop/share/hadoop/
yarn/hadoop-yarn-applications-distributedshell-2.6.0.
jar:/usr/local/hadoop/share/hadoop/yarn/hadoop-yarn-com-
mon-2.6.0.jar:/usr/local/hadoop/share/hadoop/yarn/hadoop-
yarn-applications-unmanaged-am-launcher-2.6.0.jar:/usr/
local/hadoop/share/hadoop/mapreduce/lib/netty-
3.6.2.Final.jar:/usr/local/hadoop/share/hadoop/mapreduce/
lib/jackson-core-asl-1.9.13.jar:/usr/local/hadoop/share/
hadoop/mapreduce/lib/snappy-java-1.0.4.1.jar:/usr/local/
hadoop/share/hadoop/mapreduce/lib/hadoop-annotations-
2.6.0.jar:/usr/local/hadoop/share/hadoop/mapreduce/lib/
asm-3.2.jar:/usr/local/hadoop/share/hadoop/mapreduce/lib/
jersey-core-1.9.jar:/usr/local/hadoop/share/hadoop/mapre-
duce/lib/jersey-server-1.9.jar:/usr/local/hadoop/share/
hadoop/mapreduce/lib/aopalliance-1.0.jar:/usr/local/ha-
doop/share/hadoop/mapreduce/lib/jersey-guice-1.9.jar:/
usr/local/hadoop/share/hadoop/mapreduce/lib/javax.inject-
1.jar:/usr/local/hadoop/share/hadoop/mapreduce/lib/junit-
4.11.jar:/usr/local/hadoop/share/hadoop/mapreduce/lib/com-
mons-compress-1.4.1.jar:/usr/local/hadoop/share/hadoop/
mapreduce/lib/leveldbjni-all-1.8.jar:/usr/local/hadoop/
share/hadoop/mapreduce/lib/hamcrest-core-1.3.jar:/usr/lo-
cal/hadoop/share/hadoop/mapreduce/lib/guice-3.0.jar:/usr/

```
local/hadoop/share/hadoop/mapreduce/lib/commons-io-2.4.jar:/usr/local/hadoop/share/hadoop/mapreduce/lib/jackson-mapper-asl-1.9.13.jar:/usr/local/hadoop/share/hadoop/mapreduce/lib/paranamer-2.3.jar:/usr/local/hadoop/share/hadoop/mapreduce/lib/protobuf-java-2.5.0.jar:/usr/local/hadoop/share/hadoop/mapreduce/lib/xz-1.0.jar:/usr/local/hadoop/share/hadoop/mapreduce/lib/avro-1.7.4.jar:/usr/local/hadoop/share/hadoop/mapreduce/lib/log4j-1.2.17.jar:/usr/local/hadoop/share/hadoop/mapreduce/lib/guice-servlet-3.0.jar:/usr/local/hadoop/share/hadoop/mapreduce/hadoop-mapreduce-client-jobclient-2.6.0.jar:/usr/local/hadoop/share/hadoop/mapreduce/hadoop-mapreduce-client-shuffle-2.6.0.jar:/usr/local/hadoop/share/hadoop/mapreduce/hadoop-mapreduce-client-common-2.6.0.jar:/usr/local/hadoop/share/hadoop/mapreduce/hadoop-mapreduce-client-hs-2.6.0.jar:/usr/local/hadoop/share/hadoop/mapreduce/hadoop-mapreduce-client-app-2.6.0.jar:/usr/local/hadoop/share/hadoop/mapreduce/hadoop-mapreduce-client-core-2.6.0.jar:/usr/local/hadoop/share/hadoop/mapreduce/hadoop-mapreduce-client-jobclient-2.6.0-tests.jar:/usr/local/hadoop/share/hadoop/mapreduce/hadoop-mapreduce-examples-2.6.0.jar:/usr/local/hadoop/share/hadoop/mapreduce/hadoop-mapreduce-client-hs-plugins-2.6.0.jar:/usr/local/hadoop/contrib/capacity-scheduler/*.jar:/usr/local/hadoop/contrib/capacity-scheduler/*.jar
13. STARTUP_MSG: build = https://git-wip-us.apache.org/repos/asf/hadoop.git -r e3496499ecb8d220fba99dc5ed4c99c8f9e33bb1; compiled by 'jenkins' on 2014-11-13T21:10Z
14. STARTUP_MSG:   java = 1.8.0_191
15. ************************************************************/
16. 20/01/13 15:08:05 INFO namenode.NameNode: registered UNIX signal handlers for [TERM, HUP, INT]
17. 20/01/13 15:08:05 INFO namenode.NameNode: createNameNode [-format]
```

18. 20/01/13 15:08:06 WARN common.Util: Path/usr/local/hadoop/hdfs/name should be specified as a URI in configuration files. Please update hdfs configuration.
19. 20/01/13 15:08:06 WARN common.Util: Path/usr/local/hadoop/hdfs/name should be specified as a URI in configuration files. Please update hdfs configuration.
20. Formatting using clusterid: CID-b453378a-dd07-42ad-b1e7-5fd533636121
21. 20/01/13 15:08:06 INFO namenode.FSNamesystem: No KeyProvider found.
22. 20/01/13 15:08:06 INFO namenode.FSNamesystem: fsLock is fair: true
23. 20/01/13 15:08:06 INFO blockmanagement.DatanodeManager: dfs.block.invalidate.limit=1000
24. 20/01/13 15:08:06 INFO blockmanagement.DatanodeManager: dfs.namenode.datanode.registration.ip-hostname-check=true
25. 20/01/13 15:08:06 INFO blockmanagement.BlockManager: dfs.namenode.startup.delay.block.deletion.sec is set to 000:00:00:00.000
26. 20/01/13 15:08:06 INFO blockmanagement.BlockManager: The block deletion will start around 2020 一月 13 15:08:06
27. 20/01/13 15:08:06 INFO util.GSet: Computing capacity for map BlocksMap
28. 20/01/13 15:08:06 INFO util.GSet: VM type   =64-bit
29. 20/01/13 15:08:06 INFO util.GSet: 2.0% max memory 889 MB = 17.8 MB
30. 20/01/13 15:08:06 INFO util.GSet: capacity   =2^21 = 2097152 entries
31. 20/01/13 15:08:06 INFO blockmanagement.BlockManager: dfs.block.access.token.enable=false
32. 20/01/13 15:08:06 INFO blockmanagement.BlockManager: defaultReplication   =3
33. 20/01/13 15:08:06 INFO blockmanagement.BlockManager: maxReplication   =512
34. 20/01/13 15:08:06 INFO blockmanagement.BlockManager: minReplication   =1

35. 20/01/13 15:08:06 INFO blockmanagement.BlockManager:maxReplicationStreams =2
36. 20/01/13 15:08:06 INFO blockmanagement.BlockManager:shouldCheckForEnoughRacks =false
37. 20/01/13 15:08:06 INFO blockmanagement.BlockManager:replicationRecheckInterval =3000
38. 20/01/13 15:08:06 INFO blockmanagement.BlockManager:encryptDataTransfer =false
39. 20/01/13 15:08:06 INFO blockmanagement.BlockManager:maxNumBlocksToLog =1000
40. 20/01/13 15:08:06 INFO namenode.FSNamesystem:fsOwner =root (auth:SIMPLE)
41. 20/01/13 15:08:06 INFO namenode.FSNamesystem:supergroup =supergroup
42. 20/01/13 15:08:06 INFO namenode.FSNamesystem:isPermissionEnabled=true
43. 20/01/13 15:08:06 INFO namenode.FSNamesystem:HA Enabled:false
44. 20/01/13 15:08:06 INFO namenode.FSNamesystem:Append Enabled:true
45. 20/01/13 15:08:06 INFO util.GSet:Computing capacity for map INodeMap
46. 20/01/13 15:08:06 INFO util.GSet:VM type =64-bit
47. 20/01/13 15:08:06 INFO util.GSet:1.0% max memory 889 MB=8.9 MB
48. 20/01/13 15:08:06 INFO util.GSet:capacity =2^20 =1048576 entries
49. 20/01/13 15:08:06 INFO namenode.NameNode:Caching file names occuring more than 10 times
50. 20/01/13 15:08:06 INFO util.GSet:Computing capacity for map cachedBlocks
51. 20/01/13 15:08:06 INFO util.GSet:VM type =64-bit
52. 20/01/13 15:08:06 INFO util.GSet:0.25% max memory 889 MB = 2.2 MB
53. 20/01/13 15:08:06 INFO util.GSet:capacity =2^18 =262144 entries
54. 20/01/13 15:08:06 INFO namenode.FSNamesystem:dfs.namenode.safemode.threshold-pct=0.9990000128746033

```
20/01/13 15:08:06 INFO namenode.FSNamesystem: dfs.namenode.safemode.min.datanodes =0
20/01/13 15:08:06 INFO namenode.FSNamesystem: dfs.namenode.safemode.extension     =30000
20/01/13 15:08:06 INFO namenode.FSNamesystem:Retry cache on namenode is enabled
20/01/13 15:08:06 INFO namenode.FSNamesystem:Retry cache will use 0.03 of total heap and retry cache entry expiry time is 600000 millis
20/01/13 15:08:06 INFO util.GSet:Computing capacity for map NameNodeRetryCache
20/01/13 15:08:06 INFO util.GSet:VM type       =64 -bit
20/01/13 15:08:06 INFO util.GSet:0.029999999329447746% max memory 889 MB =273.1 KB
20/01/13 15:08:06 INFO util.GSet:capacity      =2^15 =32768 entries
20/01/13 15:08:06 INFO namenode.NNConf:ACLs enabled? false
20/01/13 15:08:06 INFO namenode.NNConf:XAttrs enabled? true
20/01/13 15:08:06 INFO namenode.NNConf:Maximum size of an xattr:16384
20/01/13 15:08:07 INFO namenode.FSImage:Allocated new BlockPoolId:BP -1430064102 -192.168.3.190 -1547449687003
20/01/13 15:08:07 INFO common.Storage:Storage directory/usr/local/hadoop/hdfs/name has been successfully formatted.
20/01/13 15:08:07 INFO namenode.NNStorageRetentionManager:Going to retain 1 images with txid > =0
20/01/13 15:08:07 INFO util.ExitUtil:Exiting with status 0
20/01/13 15:08:07 INFO namenode.NameNode:SHUTDOWN_MSG:
/************************************************************
SHUTDOWN_MSG:Shutting down NameNode at YYR/192.168.3.190
************************************************************/
[root@YYR ~]#
```

4. 配置各从节点 Hadoop 运行参数

由于各从节点的 Hadoop 运行参数与主节点的 Hadoop 运行参数配置一致，所以还是采用

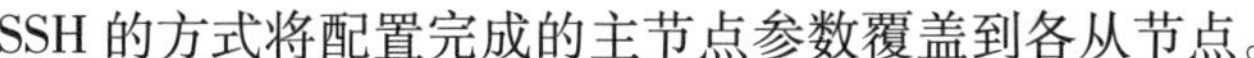

SSH 的方式将配置完成的主节点参数覆盖到各从节点。

```
[root@YYR ~]# scp /usr/local/hadoop/etc/hadoop/slaves root@Superintendent1:/usr/local/hadoop/etc/hadoop/slaves
slaves                                100%   31     1.6KB/s   00:00
[root@YYR ~]# scp /usr/local/hadoop/etc/hadoop/slaves root@Superintendent2:/usr/local/hadoop/etc/hadoop/slaves
slaves                                100%   31     2.2KB/s   00:00

[root@YYR ~]# scp /usr/local/hadoop/etc/hadoop/core-site.xml root@Superintendent1:/usr/local/hadoop/etc/hadoop/core-site.xml
core-site.xml                         100% 1195    68.2KB/s   00:00
[root@YYR ~]# scp /usr/local/hadoop/etc/hadoop/core-site.xml root@Superintendent2:/usr/local/hadoop/etc/hadoop/core-site.xml
core-site.xml                         100% 1195     2.3KB/s   00:00

[root@YYR ~]# scp /usr/local/hadoop/etc/hadoop/hdfs-site.xml root@Superintendent1:/usr/local/hadoop/etc/hadoop/hdfs-site.xml
hdfs-site.xml                         100% 1356    36.1KB/s   00:00
[root@YYR ~]# scp /usr/local/hadoop/etc/hadoop/hdfs-site.xml root@Superintendent2:/usr/local/hadoop/etc/hadoop/hdfs-site.xml
hdfs-site.xml                         100% 1356    75.2KB/s   00:00

[root@YYR ~]# scp /usr/local/hadoop/etc/hadoop/mapred-site.xml root@Superintendent1:/usr/local/hadoop/etc/hadoop/mapred-site.xml
mapred-site.xml                                                  100%
 915    79.3KB/s   00:00
[root@YYR ~]# scp /usr/local/hadoop/etc/hadoop/mapred-site.xml root@Superintendent2:/usr/local/hadoop/etc/hadoop/mapred-site.xml
mapred-site.xml                       100%  915   917.7KB/s   00:00

```

```
[root@YYR ~]# scp /usr/local/hadoop/etc/hadoop/yarn-site.xml root@Superintendent1:/usr/local/hadoop/etc/hadoop/yarn-site.xml
yarn-site.xml                       100% 1045   113.8KB/s   00:00
[root@YYR ~]# scp /usr/local/hadoop/etc/hadoop/yarn-site.xml root@Superintendent2:/usr/local/hadoop/etc/hadoop/yarn-site.xml
yarn-site.xml                       100% 1045    54.3KB/s   00:00

[root@YYR ~]#
```

5. 初次启动 Hadoop

```
# 将当前路径切换到/usr/local/hadoop/sbin/目录
[root@YYR ~]# cd /usr/local/hadoop/sbin/

# 运行 Hadoop 启动脚本
[root@YYR sbin]# start-all.sh
This script is Deprecated. Instead use start-dfs.sh and start-yarn.sh
Starting namenodes on [YYR]
The authenticity of host 'YYR (192.168.3.190)' can't be established.
ECDSA key fingerprint is SHA256:kRhfH0fSnLfPXr70iC2QIiPaQJ2S8Hi3NoMRlB0Jm80.
ECDSA key fingerprint is MD5:ba:ec:f6:61:68:ac:c9:a2:f3:06:c5:6c:15:b1:e5:7f.
# 输入yes,按 Enter 键
Are you sure you want to continue connecting (yes/no)? yes
YYR: Warning: Permanently added 'YYR,192.168.3.190' (ECDSA) to the list of known hosts.
YYR: starting namenode, logging to /usr/local/hadoop/logs/hadoop-root-namenode-YYR.out
YYR: starting datanode, logging to /usr/local/hadoop/logs/hadoop-root-datanode-YYR.out
Superintendent1: starting datanode, logging to /usr/local/hadoop/logs/hadoop-root-datanode-Superintendent1.out
```

```
Superintendent2:starting datanode, logging to/usr/local/hadoop/logs/hadoop-root-datanode-Superintendent2.out
Starting secondary namenodes[YYR]
YYR:starting secondarynamenode, logging to/usr/local/hadoop/logs/hadoop-root-secondarynamenode-YYR.out
starting yarn daemons
starting resourcemanager, logging to/usr/local/hadoop/logs/yarn-root-resourcemanager-YYR.out
Superintendent2: starting nodemanager, logging to/usr/local/hadoop/logs/yarn-root-nodemanager-Superintendent2.out
YYR:starting nodemanager, logging to/usr/local/hadoop/logs/yarn-root-nodemanager-YYR.out
Superintendent1: starting nodemanager, logging to/usr/local/hadoop/logs/yarn-root-nodemanager-Superintendent1.out
[root@YYR sbin]#
```

至此，Hadoop 完全分布式安装部署的工作全部完成。下面将对 Hadoop 集群进行验证。

任务6 验证 Hadoop

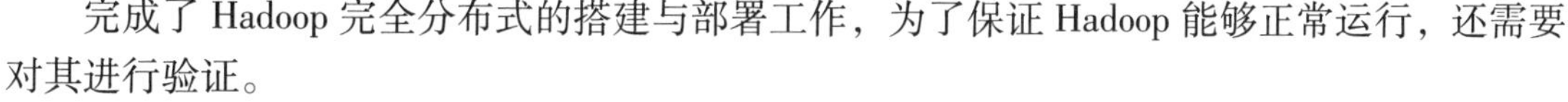

完成了 Hadoop 完全分布式的搭建与部署工作，为了保证 Hadoop 能够正常运行，还需要对其进行验证。

1. 启动 Hadoop

在没有进行任何操作前，Hadoop 不会随着运行平台启动，需要对平台进行启动验证。

```
# 切换 Hadoop 运行目录
[root@YYR ~]# cd/usr/local/hadoop/sbin/

# 启动 Hadoop 集群
[root@YYR sbin]# start-all.sh

This script is Deprecated. Instead use start-dfs.sh and start-yarn.sh
Starting namenodes on[YYR]
YYR:starting namenode, logging to/usr/local/hadoop/logs/hadoop-root-namenode-YYR.out
YYR:starting datanode, logging to/usr/local/hadoop/logs/hadoop-root-datanode-YYR.out
```

```
Superintendent2: starting datanode, logging to /usr/local/hadoop/logs/hadoop-root-datanode-Superintendent2.out
Superintendent1: starting datanode, logging to /usr/local/hadoop/logs/hadoop-root-datanode-Superintendent1.out
Starting secondary namenodes [YYR]
YYR: starting secondarynamenode, logging to /usr/local/hadoop/logs/hadoop-root-secondarynamenode-YYR.out
starting yarn daemons
starting resourcemanager, logging to /usr/local/hadoop/logs/yarn-root-resourcemanager-YYR.out
YYR: starting nodemanager, logging to /usr/local/hadoop/logs/yarn-root-nodemanager-YYR.out
Superintendent1: starting nodemanager, logging to /usr/local/hadoop/logs/yarn-root-nodemanager-Superintendent1.out
Superintendent2: starting nodemanager, logging to /usr/local/hadoop/logs/yarn-root-nodemanager-Superintendent2.out
[root@YYR sbin]#
```

2. 验证 Hadoop 进程

在集群内，全部主机节点运行 jps 命令查看 Hadoop 进程运行情况，Hadoop 各节点显示如下结果表示运行正常。

（1）YYR 节点

```
[root@YYR sbin]# jps
1489 NameNode
1937 ResourceManager
2340 Jps
2037 NodeManager
1770 SecondaryNameNode
1613 DataNode
[root@YYR sbin]#
```

（2）Superintendent1 节点

```
[root@Superintendent1 ~]# jps
1360 DataNode
1457 NodeManager
1586 Jps
[root@Superintendent1 ~]#
```

（3）Superintendent2 节点

```
1.[root@Superintendent2 ~]# jps
2.1600 Jps
3.1372 DataNode
4.1469 NodeManager
5.[root@Superintendent2 ~]#
```

3. 验证 Web 页面访问

验证 WEB 页面访问

（1）HDFS 管理界面

在浏览器地址栏输入“192.168.3.190:50070”，登录后，“Overview”和“Datanode”界面分别如图 2－3 和图 2－4 所示。

Hadoop Overview Datanodes Snapshot Startup Progress Utilities

Overview 'Master:9000' (active)

Started:	Mon Jan 13 17:43:08 CST 2020
Version:	2.6.0, re3496499ecb8d220fba99dc5ed4c99c8f9e33bb1
Compiled:	2014-11-13T21:10Z by jenkins from (detached from e349649)
Cluster ID:	CID-f7c837a9-fb48-4f3f-bd3c-bfd53155997a
Block Pool ID:	BP-1474765124-192.168.3.190-1547458886258

Summary

Security is off.

Safemode is off.

1 files and directories, 0 blocks = 1 total filesystem object(s).

Heap Memory used 95.8 MB of 226.5 MB Heap Memory. Max Heap Memory is 889 MB.

Non Heap Memory used 42.29 MB of 43.28 MB Commited Non Heap Memory. Max Non Heap Memory is -1 B.

Configured Capacity:	261.75 GB
DFS Used:	12 KB
Non DFS Used:	5.04 GB
DFS Remaining:	256.71 GB
DFS Used%:	0%
DFS Remaining%:	98.07%
Block Pool Used:	12 KB
Block Pool Used%:	0%
DataNodes usages% (Min/Median/Max/stdDev):	0.00% / 0.00% / 0.00% / 0.00%

图 2－3 HDFS 管理界面概览

Hadoop Overview Datanodes Snapshot Startup Progress Utilities

Datanode Information

In operation

Node	Last contact	Admin State	Capacity	Used	Non DFS Used	Remaining	Blocks	Block pool used	Failed Volumes	Version
Master (192.168.3.190:50010)	2	In Service	81.84 GB	4 KB	1.92 GB	79.92 GB	0	4 KB (0%)	0	2.6.0
Slave1 (192.168.3.191:50010)	0	In Service	89.96 GB	4 KB	1.56 GB	88.4 GB	0	4 KB (0%)	0	2.6.0
Slave2 (192.168.3.192:50010)	0	In Service	89.96 GB	4 KB	1.56 GB	88.4 GB	0	4 KB (0%)	0	2.6.0

Decomissioning

Node	Last contact	Under replicated blocks	Blocks with no live replicas	Under Replicated Blocks In files under construction

Hadoop, 2014.

图 2－4 HDFS 通信界面

(2) YARN 管理界面

在浏览器地址栏输入“192.168.3.190:8088”，显示如图 2-5 所示界面。

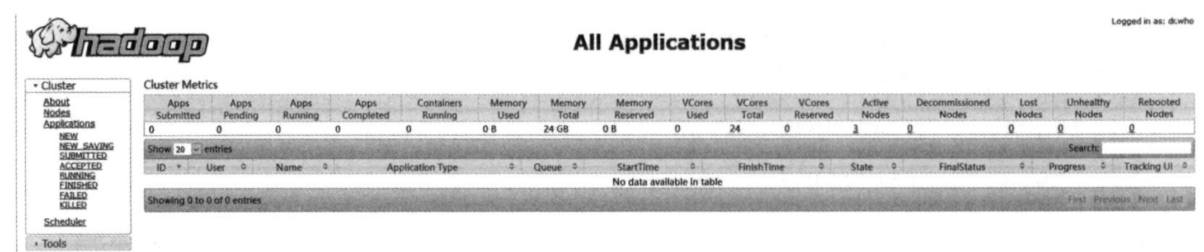

图 2-5　YARN 管理界面

4. 验证集群功能

Hadoop 集群理论上已经搭建成功，使用前面的验证方法进行检查，发现 HDFS、YARN 都处于正常运行的状态，下面需要对搭建好的 Hadoop 集群做功能上的验证。Hadoop 软件包提供了官方的测试程序，该程序存放在以下路径：

/usr/local/hadoop/share/hadoop/mapreduce/hadoop-mapreduce-examples-2.6.0.jar

(1) HDFS 功能验证

在 Hadoop 软件包部署路径下存放了一个 README.txt，将该文档上传到 HDFS 中。同时，在 HDFS 中创建一系列目录，为 MapReduce 的运行做准备。

HDFS 功能验证

```
# 在 HDFS 根目录下创建目录 testdata
[root@YYR ~]# hadoop fs -mkdir /testdata

# 在 testdata 目录下创建子目录 input,用于上传数据文件
[root@YYR ~]# hadoop fs -mkdir /testdata/input

# 将数据文件 /usr/local/hadoop/README.txt 上传到 HDFS 的 /testdata/input 目录下
[root@YYR ~]# hadoop fs -put /usr/local/hadoop/README.txt /testdata/input

# 查看文件上传结果
[root@YYR ~]# hadoop fs -ls /testdata/input
Found 1 items
-rw-r--r--3 root supergroup        1366 2020-01-15 09:20 /testdata/input/README.txt
```

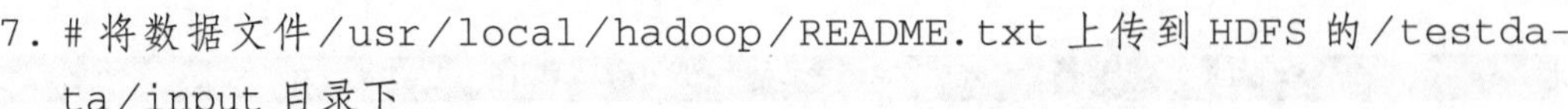

```
 
#创建/testdata/output,用于存放MapReduce计算后的结果文件
[root@YYR ~]# hadoop fs -mkdir /testdata/output
 
[root@YYR ~]#
```

经过上述步骤的测试，认为Hadoop集群中的HDFS可以正常运行。

（2）MapReduce功能验证

调用Hadoop软件包中提供的WordCount验证程序来对README.txt中的单词进行统计，以验证Hadoop集群的MapReduce组件功能正常。

```
#运行wordcount示例程序,以/testdata/input作为输入,将计算结果输出到/testdata/output/result目录中
[root@YYR sbin]# hadoop jar /usr/local/hadoop/share/hadoop/mapreduce/ hadoop-mapreduce-examples-2.6.0.jar wordcount /testdata/input /testdata/output/result
20/01/15 10:06:24 INFO client.RMProxy:Connecting to ResourceManager at /0.0.0.0:8032
20/01/15 10:06:25 INFO input.FileInputFormat:Total input paths to process:1
20/01/15 10:06:25 INFO mapreduce.JobSubmitter: number of splits:1
20/01/15 10:06:26 INFO mapreduce.JobSubmitter:Submitting tokens for job:job_1547517519741_0001
20/01/15 10:06:26 INFO impl.YarnClientImpl:Submitted application application_1547517519741_0001
20/01/15 10:06:26 INFO mapreduce.Job:The url to track the job:http://YYR:8088/proxy/application_1547517519741_0001/
20/01/15 10:06:26 INFO mapreduce.Job: Running job: job_1547517519741_0001
20/01/15 10:06:33 INFO mapreduce.Job:Job job_1547517519741_0001 running in uber mode:false
20/01/15 10:06:33 INFO mapreduce.Job:map 0% reduce 0%
20/01/15 10:06:38 INFO mapreduce.Job:map 100% reduce 0%
20/01/15 10:06:43 INFO mapreduce.Job:map 100% reduce 100%
20/01/15 10:06:43 INFO mapreduce.Job:Job job_1547517519741_0001 completed successfully
20/01/15 10:06:43 INFO mapreduce.Job:Counters:49
```

```
    File System Counters
        FILE:Number of bytes read=1836
        FILE:Number of bytes written=214743
        FILE:Number of read operations=0
        FILE:Number of large read operations=0
        FILE:Number of write operations=0
        HDFS:Number of bytes read=1475
        HDFS:Number of bytes written=1306
        HDFS:Number of read operations=6
        HDFS:Number of large read operations=0
        HDFS:Number of write operations=2
    Job Counters
        Launched map tasks=1
        Launched reduce tasks=1
        Data-local map tasks=1
        Total time spent by all maps in occupied slots (ms)=2790
        Total time spent by all reduces in occupied slots (ms)=2891
        Total time spent by all map tasks (ms)=2790
        Total time spent by all reduce tasks (ms)=2891
        Total vcore-seconds taken by all map tasks=2790
        Total vcore-seconds taken by all reduce tasks=2891
        Total megabyte-seconds taken by all map tasks=2856960
        Total megabyte-seconds taken by all reduce tasks=2960384
    Map-Reduce Framework
        Map input records=31
        Map output records=179
        Map output bytes=2055
        Map output materialized bytes=1836
        Input split bytes=109
        Combine input records=179
        Combine output records=131
        Reduce input groups=131
```

```
                Reduce shuffle bytes=1836
                Reduce input records=131
                Reduce output records=131
                Spilled Records=262
                Shuffled Maps =1
                Failed Shuffles=0
                Merged Map outputs=1
                GC time elapsed (ms)=317
                CPU time spent (ms)=1840
                Physical memory (bytes) snapshot=432300032
                Virtual memory (bytes) snapshot=4261875712
                Total committed heap usage (bytes)=347602944
        Shuffle Errors
                BAD_ID=0
                CONNECTION=0
                IO_ERROR=0
                WRONG_LENGTH=0
                WRONG_MAP=0
                WRONG_REDUCE=0
        File Input Format Counters
                Bytes Read=1366
        File Output Format Counters
                Bytes Written=1306

# 查看计算结果文件生成情况
[root@YYR sbin]# hadoop fs -ls /testdata/output/result
Found 2 items
-rw-r--r--   3 root supergroup          0 2020-01-15 10:06 /testdata/output/result/_SUCCESS
-rw-r--r--   3 root supergroup       1306 2020-01-15 10:06 /testdata/output/result/part-r-00000

# 查看运行结果
[root@YYR sbin]# hadoop fs -cat /testdata/output/result/part-r-00000
(BIS),1
(ECCN)  1
```

```
(TSU)	1
(see	1
5D002.C.1,	1
740.13)	1
<http://www.wassenaar.org/>	1
Administration	1
Apache	1
BEFORE	1
BIS	1
Bureau	1
Commerce,	1
Commodity	1
Control	1
Core	1
Department	1
ENC	1
Exception	1
Export	2
For	1
Foundation	1
Government	1
Hadoop	1
Hadoop,	1
Industry	1
Jetty	1
License	1
Number	1
Regulations,	1
SSL	1
Section	1
Security	1
See	1
Software	2
Technology	1
The	4
This	1
U.S.	1
```

```
Unrestricted    1
about   1
algorithms.     1
and     6
and/or  1
anothcr 1
any     1
as      1
asymmetric      1
at:2
both    1
by      1
check   1
classified      1
code    1
code.   1
concerning      1
country 1
country's       1
country,        1
cryptographic   3
currently       1
details 1
distribution    2
eligible        1
encryption      3
exception       1
export  1
following       1
for     3
form    1
from    1
functions       1
has     1
have    1
http://hadoop.apache.org/core/1
http://wiki.apache.org/hadoop/1
```

```
156.if      1
157.import,2
158.in      1
159.included        1
160.includes        2
161.information     2
162.information.     1
163.is      1
164.it      1
165.latest  1
166.laws,  1
167.libraries       1
168.makes   1
169.manner  1
170.may     1
171.more    2
172.mortbay.org.     1
173.object  1
174.of      5
175.on      2
176.or      2
177.our     2
178.performing      1
179.permitted.      1
180.please  2
181.policies        1
182.possession,    2
183.project 1
184.provides        1
185.re-export       2
186.regulations     1
187.reside  1
188.restrictions     1
189.security        1
190.see     1
191.software        2
192.software,    2
```

```
software.	2
software:	1
source	1
the	8
this	3
to	2
under	1
use,	2
uses	1
using	2
visit	1
website	1
which	2
wiki,	1
with	1
written	1
you	1
your	1
[root@YYR sbin]#
```

根据上述步骤，MapReduce 组件运行正常。

学习笔记

项目三
数据清洗技术

项目描述

在完成了 Hadoop 离线环境的部署后，我们将站在数据清洗人员的角度来了解数据清洗（Data Cleaning）所要具备的知识。

数据清洗是发现并纠正数据错误的最后一道程序，包括检查数据一致性、处理无效值和缺失值等。在对数据进行重新审查和校验的过程中，需要删除重复的数据、纠正错误的数据、转换需求不一致的数据。因为数据是通过数据爬虫组从多个外部网站上抓取过来的，这样就避免不了有的数据是错误数据、有的数据相互之间有冲突的问题，这些错误的或者有冲突的数据显然是我们不想要的，称为“脏数据”。需要按照项目需求把“脏数据”洗掉，这就是数据清洗。

数据清洗的流程如图 3－1 所示。

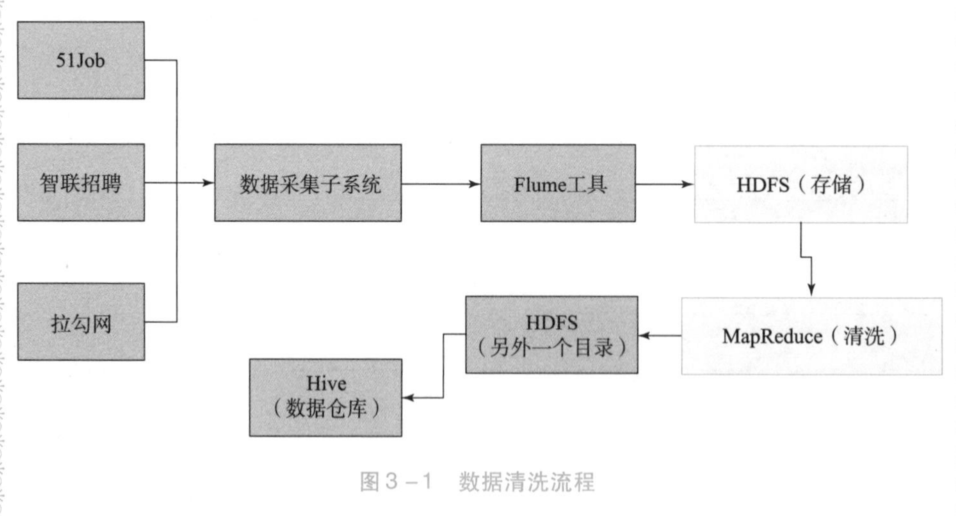

图 3－1　数据清洗流程

项目分析

作为数据清洗工作人员，由流程图发现：完成数据清洗工作必须依赖HDFS的知识和MapReduce的知识。本章将完成如下任务：

任务1：HDFS分布式文件系统的体系架构和文件操作。

任务2：MapReduce分布式计算系统的基本运行框架。

任务3：YARN分布式资源管理平台的资源调度。

任务1 HDFS分布式文件系统的体系架构和文件操作

HDFS（Hadoop Distributed File System）是一个类似于GFS的开源的分布式文件系统。它提供了一个可扩展、高可靠、高可用的大规模数据分布式存储管理系统，基于物理上分布在各个数据存储节点的本地Linux系统的文件系统，为上层应用程序提供了一个逻辑上成为整体的大规模数据存储文件系统。HDFS采用多副本（默认为3个副本）数据冗余存储机制，并提供了有效的数据出错检测和数据恢复机制，大大提高了数据存储的可靠性。首先理解几个概念：

- Block（数据块）
- NameNode（元数据节点）
- DataNode（数据节点）
- SecondaryNameNode（从元数据节点）

HDFS存储原理

1. Block（数据块）

在HDFS中，最基本的存储单位是Block。版本Hadoop 1. x中默认的Block大小是64 MB，当前是Hadoop 2. x版本，所以默认Block大小为128 MB。在HDFS中，任何数据文件都是被切分成128 MB一块的Block进行存储的，并且默认情况下，每个Block的副本是3个，存放到不同的DataNode（数据节点）中。但是，如果一个文件小于一个数据块的大小（小于128 MB），则并不占用整个数据块存储空间，大小多少则占用多少空间。假设一个文件为200 MB，那么会切成两个Block，其中第二个Block只有72 MB。

2. NameNode（元数据节点）

NameNode节点用来管理文件系统的命名空间，并且接受客户端的读写请求。在该节点中有两个文件（FsImage和Edits）用来存储元数据。其中，Edits文件存放对元数据操作的日志。但是HDFS上的每个数据文件包括哪些数据块、分布在哪些数据节点上，这些信息是在系统启动时从DataNode（数据节点）收集而成的。

元数据（MetaData）包括：

- 文件的 OwerShip 和 Permissions
- 文件的大小和时间
- Block 列表（包括 Block 的位置信息和 Block 的偏移量）
- Block 每个副本的位置信息

NameNode 中的文件夹结构如图 3－2 所示。

```
/usr/local/hadoop/hdfs/name/current/

-rw-r--r--. 1 root root   30895 4月  15 23:29 edits_0000000000000000098-0000000000000000344
-rw-r--r--. 1 root root      42 4月  16 00:29 edits_0000000000000000345-0000000000000000346
-rw-r--r--. 1 root root      42 4月  16 01:29 edits_0000000000000000347-0000000000000000348
-rw-r--r--. 1 root root   10506 4月  16 02:29 edits_0000000000000000349-0000000000000000432
-rw-r--r--. 1 root root     182 4月  16 03:29 edits_0000000000000000433-0000000000000000436
-rw-r--r--. 1 root root    3320 4月  16 04:29 edits_0000000000000000437-0000000000000000465
-rw-r--r--. 1 root root 1048576 4月  16 04:29 edits_inprogress_0000000000000000466
-rw-r--r--. 1 root root    5425 4月  16 03:29 fsimage_0000000000000000436
-rw-r--r--. 1 root root      62 4月  16 03:29 fsimage_0000000000000000436.md5
-rw-r--r--. 1 root root    5425 4月  16 04:29 fsimage_0000000000000000465
-rw-r--r--. 1 root root      62 4月  16 04:29 fsimage_0000000000000000465.md5
-rw-r--r--. 1 root root       4 4月  16 04:29 seen_txid
-rw-r--r--. 1 root root     207 4月  15 22:36 VERSION
```

图 3－2　NameNode 中的文件夹结构

VERSION 保存了 HDFS 数据的版本信息。

其中：

➢ layoutVersion 保存了 HDFS 硬盘上数据结构的格式版本号。

➢ namespaceID 是文件系统的唯一标识符，是在文件系统初次格式化时生成的。

➢ storageType 表示此文件夹中保存的是元数据节点的数据结构。

3. DataNode（数据节点）

DATANODE（数据节点）

DataNode 是 HDFS 中真正存放数据的地方，这里存放的就是数据文件的内容。客户端（Client）或者 NameNode 可以向 DataNode 节点请求写入或者读取数据块。并且 DataNode 会周期性地向 NameNode 节点汇报其存储的 Block 信息，和 NameNode 一直做到持续连接。

4. SecondaryNameNode（从元数据节点）

SecondaryNameNode 节点并不是 NameNode 节点出现问题时的备用节点，它和 NameNode 节点负责不同的事情。其主要功能就是周期性地将 NameNode 节点上的 FsImage 和 Edits 日志文件合并（也叫 chkpoint 操作），以防止 Edits 日志文件过大。合并过后得到一个新的 FsImage 文件复制给 NameNode，并且 SecondaryNameNode 节点上也保存了一份，万一 NameNode 上的元数据丢失，还可以恢复部分元数据，如图 3－3 所示。

简单了解了 HDFS 中几个重要概念后，我们思考一下：一个 Hadoop 集群到底应该有多大？实际上这个问题没有标准答案。Hadoop 最大的优势在于用户可以在初始阶段构建一个小集群（大约 10 个节点），并随着存储与计算需求增长继续扩充。或者可以理解为，您的集群需要增长得多快？HDFS 的体系架构就可以按照您的需求增加，如图 3－4 所示。

HDFS 是一个 YYR/Slave 的架构，一个 YYR 会带 N 个 slave。这在大数据框架里是很常

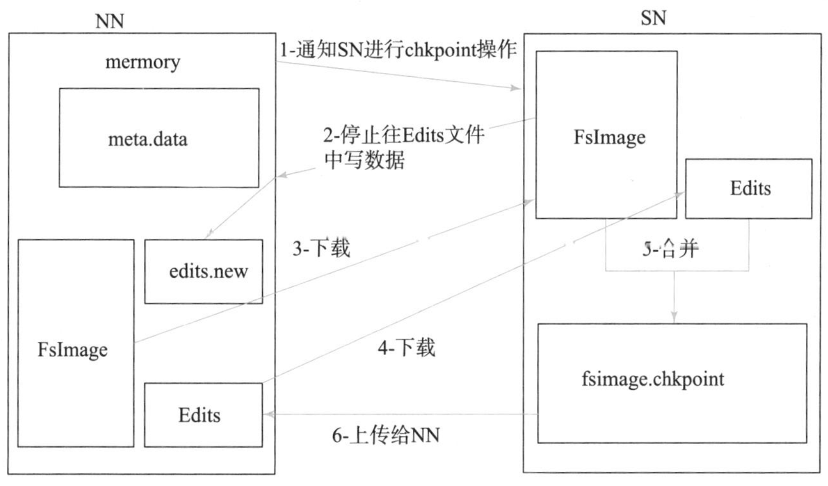

图 3-3　FsImage 文件和 Edits 日志文件的合并图

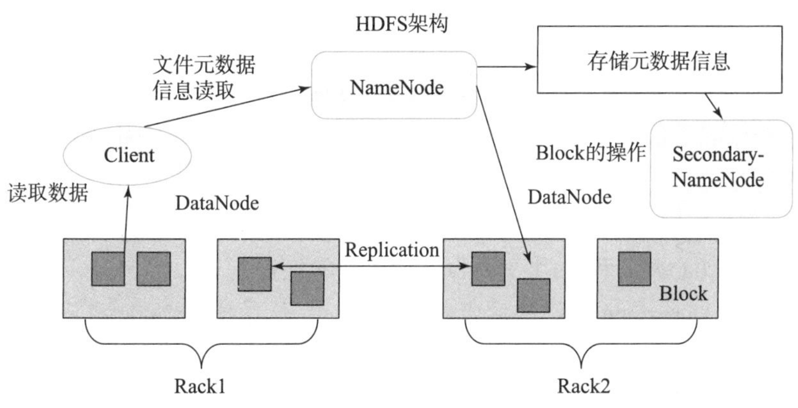

图 3-4　HDFS 体系架构

见的，包括后面要掌握的 YARN 和 HBase。

一个 HDFS 的集群通常由一个 NameNode 和多个 DataNodes 构成。YYR 就是 NameNode，slave 就是 DataNode。

➢ YYR 的主要功能

①管理和存储文件系统的元数据。

②接受客户端访问文件系统里的文件。

➢ Slave 的主要功能

①管理文件存储到节点上。

②定期向 NameNode 发送心跳信息，汇报本身及其存储的 Block 信息、健康状况。

③根据客户端的请求，读和写 Block。

④负责 Block 的创建、删除和备份工作。

⑤一个文件会被拆分成多个 Block，每个 Block 有一个 BlockSize 属性，默认是 128 MB。

➢ Client 客户端

①切分文件的 Block。

②负责读和写数据文件。

③有 HDFS 的 Shell 客户端和 Java API 的客户端两种。

NameNode 和 DataNode 希望能够在一些非常廉价的机器上运行，并且这些机器都是 Linux 系统。

HDFS 的开发构建都是基于 Java 语言的，因此，任何能够运行 Java 的机器，都能够运行 NameNode 或者是 DataNode 的软件。

典型的架构就是，一台机器上面运行一个 NameNode，其他机器可以选择一台机器运行一个 DataNode，或一台机器运行多个 DataNode，但是在实际生产环境中，不建议一台机器运行多个 DataNode。

HDFS 是一个高可靠的文件，并且这些文件是跨节点的。在一个集群中，一个文件会被分割成很多 Block 存储在节点中。一个文件会被拆分成一个有序的 Block。这些 Block 都是以多副本的方式存储的，为了容错。因为如果只存储在一个节点上，如果那个节点挂掉，就没办法访问了。如果是多副本存储，即使其中的几个节点不能访问，其他节点依然能正常访问 Block 的副本。

Block 的大小和 Replication factor（副本系数）都可以针对一个文件来单独配置。

一个文件里所有的 Block，除了最后一个外，其他的大小都是一样的。

一个应用程序能够指定一个文件的 Replication factor，可以在创建文件的时候指定，也可以在后面进行相应的修改。

NameNode 还要决定所有 Block 的副本操作，它会周期性地从集群中的每个 DataNode 处接收一个 Heartbeat（心跳）和 一个 Blockreport（块的汇报）。

Heartbeat 告诉 NameNode 自己还活着，Blockreport 告诉 NameNode 自己机器上块的一些情况。

HDFS 是一个高可靠的文件，并且这些文件是跨节点的。在一个集群中，一个文件会被分割成很多 Block 存储在节点中。一个文件会被拆分成一个有序的 Block。这些 Block 都是以多副本的方式存储的。因为如果只存储在一个节点上，当那个节点挂掉时，就不能访问了。如果是多副本存储，即使其中的几个节点不能访问，其他节点依然能正常访问 Block 的副本。Block 的大小在该段后添加，如图 3 -5 所示。

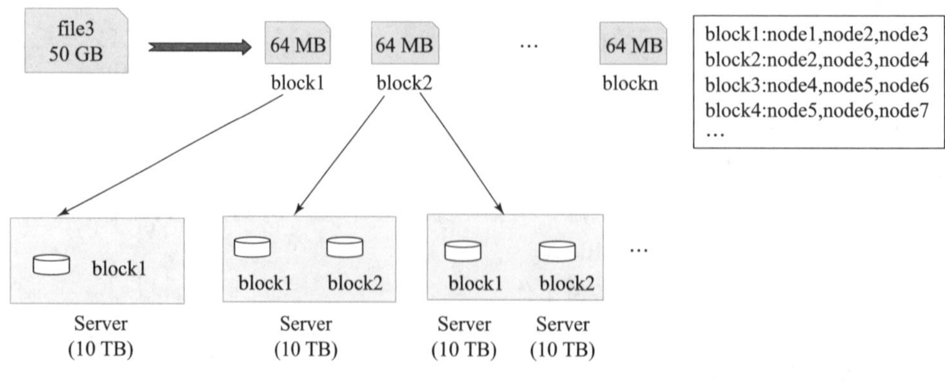

图 3 -5　Block 在 HDFS 中的存储

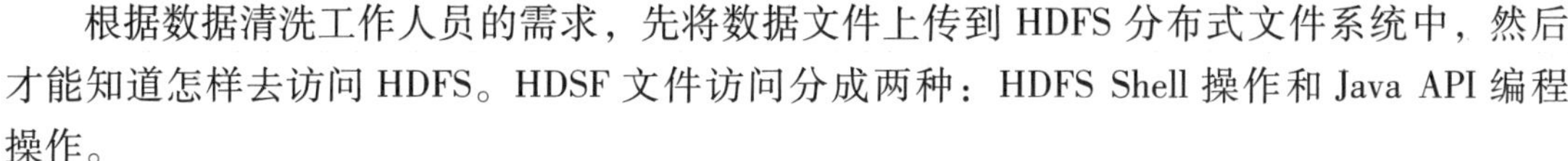

根据数据清洗工作人员的需求，先将数据文件上传到 HDFS 分布式文件系统中，然后才能知道怎样去访问 HDFS。HDSF 文件访问分成两种：HDFS Shell 操作和 Java API 编程操作。

1. HDFS Shell 操作文件

HDFS Shell 操作通过命令行对数据文件进行管理。

```
[root@YYR ~]# hdfs dfs -help
Usage:hadoop fs[generic options]
     [-appendToFile <localsrc>... <dst>]
     [-cat[-ignoreCrc] <src>...]
     [-checksum <src>...]
     [-chgrp[-R]GROUP PATH...]
     [-chmod[-R] <MODE[,MODE]... |OCTALMODE> PATH...]
     [-chown[-R][OWNER][:[GROUP]]PATH...]
     [-copyFromLocal[-f][-p][-l] <localsrc>... <dst>]
     [-copyToLocal[-p][-ignoreCrc][-crc] <src>... <localdst>]
     [-count[-q][-h] <path>...]
     [-cp[-f][-p |-p[topax]] <src>... <dst>]
     [-createSnapshot <snapshotDir>[ <snapshotName>]]
     [-deleteSnapshot <snapshotDir> <snapshotName>]
     [-df[-h][ <path>...]]
     [-du[-s][-h] <path>...]
     [-expunge]
     [-get[-p][-ignoreCrc][-crc] <src>... <localdst>]
     [-getfacl[-R] <path>]
     [-getfattr[-R]{-n name |-d}[-e en] <path>]
     [-getmerge[-nl] <src> <localdst>]
     [-help[cmd ...]]
     [-ls[-d][-h][-R][ <path>...]]
     [-mkdir[-p] <path>...]
     [-moveFromLocal <localsrc>... <dst>]
     [-moveToLocal <src> <localdst>]
     [-mv <src>... <dst>]
     [-put[-f][-p][-l] <localsrc>... <dst>]
     [-renameSnapshot <snapshotDir> <oldName> <newName>]
```

```
[ -rm[ -f][ -r|-R][ -skipTrash] <src>...]
[ -rmdir[ --ignore-fail-on-non-empty] <dir>...]
[ -setfacl[ -R][{ -b|-k}{ -m|-x <acl_spec>}<path>]|[ --set <acl_spec> <path>]]
[ -setfattr{ -n name[ -v value]|-x name}<path>]
[ -setrep[ -R][ -w] <rep> <path>...]
[ -stat[format] <path>...]
[ -tail[ -f] <file>]
[ -test -[defsz] <path>]
[ -text[ -ignoreCrc] <src>...]
[ -touchz <path>...]
[ -usage[cmd ...]]
```

HDFS SHELL 操作命令

这些就是 HDFS Shell 的全部命令。这些命令和 Linux 操作系统本身的命令非常类似，所以使用 HDFS Shell 命令时，可以参考 Linux 操作系统的基本命令。

下面根据 HDFS Shell 命令完成如下操作：

①在 HDFS 中创建一个目录：/user/test/。

②上传两个本地文件（/home/newFile、oldFile）到 HDFS 目录/user/test/。

③删除 HDFS 上/user/test/目录下的 oldFile 文件。

④查看 HDFS 上/user/test/目录下的 newFile 文件内容。

⑤下载 HDFS 上/user/test/目录下的 newFile 文件到 Linux 本地的/opt 目录下。

⑥修改 HDFS 上/user/test/目录下的 newFile 文件权限为所有用户只读。

（1）创建目录

```
[root@YYR ~]# hdfs dfs -mkdir -p /user/test/
[root@YYR ~]# hdfs dfs -ls /user
20/04/17 01:11:57 WARN util.NativeCodeLoader: Unable to load native-hadoop library for your platform... using builtin-java classes where applicable
Found 1 items
drwxr-xr-x - root supergroup      0 2020-04-17 01:08 /user/test
```

通过浏览器查看目录情况，如图 3-6 所示。

Browse Directory

/user　Go!

Permission	Owner	Group	Size	Replication	Block Size	Name
drwxr-xr-x	root	supergroup	0 B	0	0 B	test

Hadoop, 2016.

图 3-6　目录显示情况

（2）上传文件到 HDFS

```
[root@YYR ~]# hdfs dfs -put /home/newFile /user/test/
[root@YYR ~]# hdfs dfs -ls /user/test/
20/04/17 01:11:57 WARN util.NativeCodeLoader: Unable to load native-hadoop library for your platform... using builtin-java classes where applicable
Found 1 items
-rw-r--r--1 root supergroup          27 2020-04-17 01:19 /user/test/newFile
还可通过另外一个命令:
[root@YYR ~]# hdfs dfs -copyFromLocal /home/oldFile /user/test/
[root@YYR ~]# hdfs dfs -ls /user/test/
20/04/17 01:25:31 WARN util.NativeCodeLoader: Unable to load native-hadoop library for your platform... using builtin-java classes where applicable
Found 2 items
-rw-r--r--1 root supergroup          27 2020-04-17 01:19 /user/test/newFile
-rw-r--r--1 root supergroup           0 2020-04-17 01:25 /user/test/oldFile
```

（3）删除 HDFS 上的文件

```
[root@YYR ~]# hdfs dfs -rm /user/test/oldFile
20/04/17 01:26:46 WARN util.NativeCodeLoader: Unable to load native-hadoop library for your platform... using builtin-java classes where applicable
20/04/17 01:26:47 INFO fs.TrashPolicyDefault: Namenode trash configuration: Deletion interval = 0 minutes, Emptier interval = 0 minutes.
Deleted /user/test/oldFile
[root@YYR ~]# hdfs dfs -ls /user/test/
20/04/17 01:28:24 WARN util.NativeCodeLoader: Unable to load native-hadoop library for your platform... using builtin-java classes where applicable
Found 1 items
-rw-r--r--1 root supergroup          27 2020-04-17 01:19 /user/test/newFile
```

(4) 查看 HDFS 目录中文件的内容

```
[root@YYR ~]# hdfs dfs -cat /user/test/newFile
20/04/17 01:29:47 WARN util.NativeCodeLoader: Unable to load native-hadoop library for your platform... using builtin-java classes where applicable
Hello Hadoop!
Hello World!
```

(5) 下载 HDFS 中的文件到本地磁盘

```
[root@YYR ~]# hdfs dfs -get /user/test/newFile /opt/
[root@YYR ~]# ls -l /opt/
总用量 8
-rw-r--r--. 1 root root   27 4月  17 01:31 newFile
drwxr-xr-x. 2 root root 4096 3月  30 03:17 test
也可以换另外一个命令:copyToLocal
[root@YYR ~]# hdfs dfs -copyToLocal /user/test/newFile /opt/
[root@YYR ~]# ls -l /opt/
总用量 8
-rw-r--r--. 1 root root   27 4月  17 01:31 newFile
drwxr-xr-x. 2 root root 4096 3月  30 03:17 test

```

(6) 修改 HDFS 上的文件权限

```
[root@YYR ~]# hdfs dfs -ls /user/test/
20/04/17 01:36:59 WARN util.NativeCodeLoader: Unable to load native-hadoop library for your platform... using builtin-java classes where applicable
Found 1 items
-rw-r--r--1 root supergroup          27 2020-04-17 01:19 /user/test/newFile
[root@YYR ~]# hdfs dfs -chmod -w /user/test/newFile
[root@YYR ~]# hdfs dfs -ls /user/test/
20/04/17 01:37:34 WARN util.NativeCodeLoader: Unable to load native-hadoop library for your platform... using builtin-java classes where applicable
Found 1 items
-r--r--r--1 root supergroup          27 2020-04-17 01:19 /user/test/newFile
```

2. Java API 操作 HDFS

本书中采用 Eclipse 开发工具，在 Windows 操作系统中安装好开发环境。完成上述任务操作。

创建 JAVA 项目

（1）创建 Java 项目

建立 Java 项目，导入 Hadoop 的所有 jar 包，如图 3－7 所示。

（2）编写代码类 HDFS_Test. java

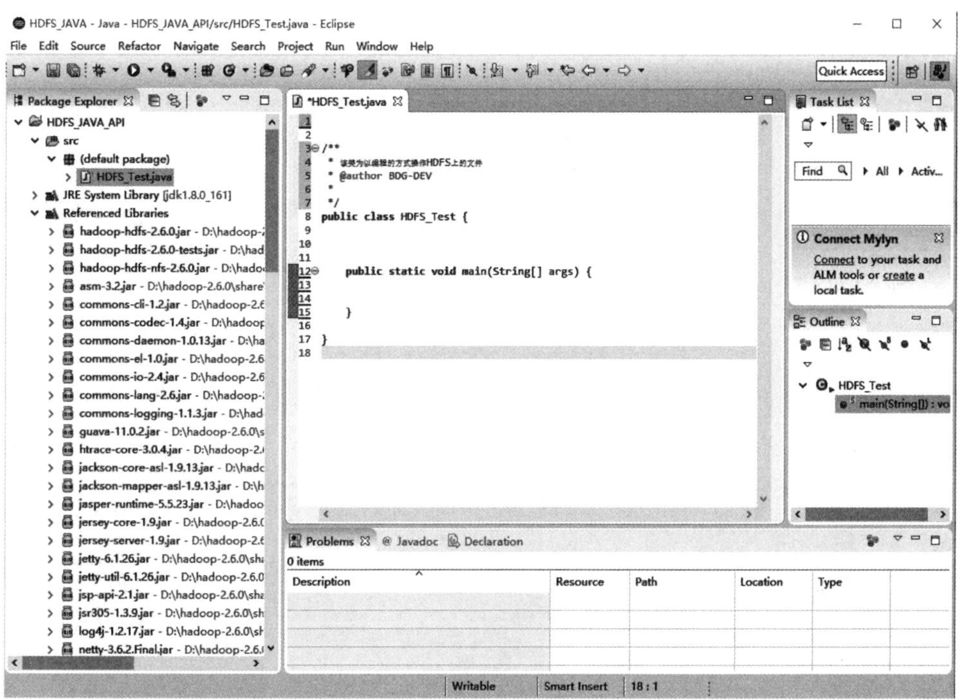

图 3－7　jar 包的导入

```
import java.io.IOException;
import java.net.URI;
import java.net.URISyntaxException;
import org.apache.hadoop.conf.Configuration;
import org.apache.hadoop.fs.FileSystem;
import org.apache.hadoop.fs.Path;
import org.apache.hadoop.fs.permission.FsAction;
import org.apache.hadoop.fs.permission.FsPermission;
public class HDFS_Test {

public static void main(String[]args) {
```

```

    HDFS_Test gfs =new HDFS_Test();

    FileSystem fs =gfs.getHadoopFS();

    gfs.createDir(fs);
    gfs.PutFileToHDFS(fs);
    gfs.getFileFromHDFS(fs);
    gfs.chmodFile(fs);
    gfs.dropHDFSPath(fs);
}

//根据配置文件获取 HDFS 操作对象
public FileSystem getHadoopFS(){

    FileSystem fs =null;

    Configuration conf =null;
    conf =new Configuration();

    String hdfsUserName ="root";
    URI hdfsUri =null;

    try{
        hdfsUri =new URI("hdfs://192.168.3.190:9000");
    } catch (URISyntaxException e) {
        //TODO Auto -generated catch block
        e.printStackTrace();
    }

    try {
        fs =FileSystem.get(hdfsUri,conf,hdfsUserName);
    } catch (IOException e) {
        //TODO Auto -generated catch block
        e.printStackTrace();
    } catch (InterruptedException e) {
        //TODO Auto -generated catch block
```

```
            e.printStackTrace();
        }
        return fs;
}

//创建一个文件

public boolean createDir(FileSystem fs){

        boolean b = false;

        Path path = new Path("/test");

        try {
            b = fs.mkdirs(path);
        } catch (IOException e) {
            //TODO Auto-generated catch block
            e.printStackTrace();
        }finally{
            try {
                fs.close();
            } catch (IOException e) {
                //TODO Auto-generated catch block
                e.printStackTrace();
            }
        }
        return b;
}


//上传文件
public void PutFileToHDFS(FileSystem fs){

        Path localPath = new Path("file:///d://demo.txt");

        Path hdfsPath = new Path("/test/");

```

```
try {
    fs.copyFromLocalFile(localPath,hdfsPath);
} catch (IOException e) {
    //TODO Auto - generated catch block
    e.printStackTrace();
}finally{
    try {
        fs.close();
    } catch (IOException e) {
        //TODO Auto - generated catch block
        e.printStackTrace();
        }
            }
        }

     //文件下载
     public void getFileFromHDFS(FileSystem fs){

         Path HDFSPath = new Path("/test/demo.txt");

         Path localPath = new Path("file:///d://test//");

         try {
             fs.copyToLocalFile(HDFSPath,localPath);
     } catch (IOException e) {
         //TODO Auto - generated catch block
         e.printStackTrace();
     }finally{
         try {
             fs.close();
         } catch (IOException e) {
             //TODO Auto - generated catch block
             e.printStackTrace();
         }
     }
 }

```

```
//修改 HDFS 上的文件权限
public void chmodFile(FileSystem fs){

    Path HDFSfile = new Path("/test/demo.txt");

     FsPermission permission = new FsPermission(FsAction.
     READ,FsAction.READ,FsAction.READ);

    try {
        fs.setPermission(HDFSfile,permission);
    } catch (IOException e) {
        //TODO Auto-generated catch block
        e.printStackTrace();
    }finally{
        try {
            fs.close();
        } catch (IOException e) {
            //TODO Auto-generated catch block
            e.printStackTrace();
        }
    }
}
//删除文件
public boolean dropHDFSPath(FileSystem fs){
    boolean b = false;

    Path path = new Path("/test");
    try {
        b = fs.delete(path,true);
    } catch (IOException e) {
        //TODO Auto-generated catch block
        e.printStackTrace();
    }finally{
        try {
            fs.close();
        } catch (IOException e) {
            //TODO Auto-generated catch block
```

```
                e.printStackTrace();
            }
        }
        return b;
    }
}
```

至此，HDFS 文件系统的操作任务完成。

任务 2　MapReduce 分布式计算系统的基本运行框架

MapReduce 是数据清洗中最重要的一个环节，由于数据采集子系统获取大量的数据，当数据量到达“大数据”这个级别的时候，必须采用一种分布式计算的方式才能完成海量数据的清洗工作，这就是 MapReduce 分布式计算技术。

MapReduce 是 Google 公司的核心计算模型，它将运行于大规模集群上的复杂的并行计算过程高度地抽象为两个函数：Map 和 Reduce。一个 MapReduce 作业（Job）通常会把输入的数据集切分为若干独立的数据块，由 Map 任务（Task）以完全并行的方式处理这些数据块。框架会先对 Map 的输出进行排序，然后把结果输入给 Reduce 任务。通常作业的输入和输出都会被存储在文件系统中。整个框架负责任务的调度和监控，以及重新执行已经失败的任务。

通常，MapReduce 框架和分布式文件系统是运行在一组相同的节点上的，也就是说，计算节点和存储节点在一起。这种配置允许框架在那些已经存好数据的节点上高效地调度任务，这样可以非常高效地利用整个集群的网络带宽。

MapReduce 被广泛地应用于日志分析、海量数据清洗、在海量数据中查找特定模式等场景中。

首先了解 MapReduce 中的几个概念：

Job：代表一个 MapReduce 程序执行时的作业。Job 可以切分成一系列运行于分布式集群中的 Map 和 Reduce 任务，每个任务只运行全部数据的一个指定的子集，以此达到整个集群的负载平衡。Map 任务通常加载、解析、转换、过滤数据，每个 Reduce 处理 Map 输出的一个子集。Reduce 任务会到 Map 任务端获取中间数据来完成分组与聚合。

Input：MapReduce 的输入是 HDFS 上存储的一系列文件集。在 Hadoop 中，这些文件被一种定义了如何分割一个文件成分片的 InputFormat 来分割。一个分片是一个文件基于字节的，可以被一个 Map 任务加载的块。

Map：每个 Map 任务被分为 Recordreader、Mapper、Combiner、Partitioner。Map 任务的输出叫作中间数据，包括 Keys 和 Values，发送到 Reduce 端。运行 Map 任务的节点尽量选择数据所在的节点，这样不需要网络传输，本地节点就可以完成计算。

➢ Recordreader

Recordreader 会把根据 InputFormat 生成的输入分片翻译成 Records。Recordreader 的目的

是把数据解析成记录，而不是解析数据本身。它把数据以键值对的形式传递给 Mapper。通常情况下，键是偏移量，值是这条记录的整个字节块。

➢ Mapper

Map 阶段，会对每个从 Recordreader 处理完的键值对执行用户代码，这些键值对又叫中间键值对。键和值的选择不是任意的，键会用来分组，值是 Reducer 端用来分析的数据。

➢ Combiner

Combiner 是一个 Map 阶段聚合数据的部件，相当于一个局部 Reducer，是个可选项。它根据用户提供的方法在一个 Mapper 范围内根据中间键去聚合值。

➢ Partitioner

Partitioner 会获取从 Mapper（或 Combiner）来的键值对，并分割成分片。每个 Reducer 一个分片。默认用哈希值，典型的使用 Md5sum。然后 Partitioner 根据 Reduce 的个数执行取余运算：Key. Hashcode( )%(Number Of Reducers)，这样能随即均匀地根据 Key 分发数据到 Reduce，但要保证不同 Mapper 的相同 Key 到同一个 Reduce。Partitioner 可以自定义，使用更高级的样式，例如排序。

Reduce：Reduce 阶段包括 OutputFormat。

Reduce 任务会把分组的数据作为输入，并对每个 Key 组执行 Reduce 方法代码。方法会传递 Key 和相关的所有值的迭代集合。很多的处理会在这个方法里执行。一旦 Reduce 方法完成，会发送 0 或多个键值对到 OutputFormat。跟 Map 一样，不同的 Reduce 依据不同的逻辑情形而不同。

OutputFormat 会把 Reduce 阶段的输出键值对根据 RecordWriter 写到文件里。默认用 Tab 键分割键值对，用 Enter 键分割不同行。这里也可以自定义为更丰富的输出格式。最后，数据被写到 HDFS，可以自定义输出格式。

来看个简单的例子：我需要统计我的目标数据中每个单词出现的次数。写好代码并提交到服务器运行，执行流程如图 3 - 8 所示。

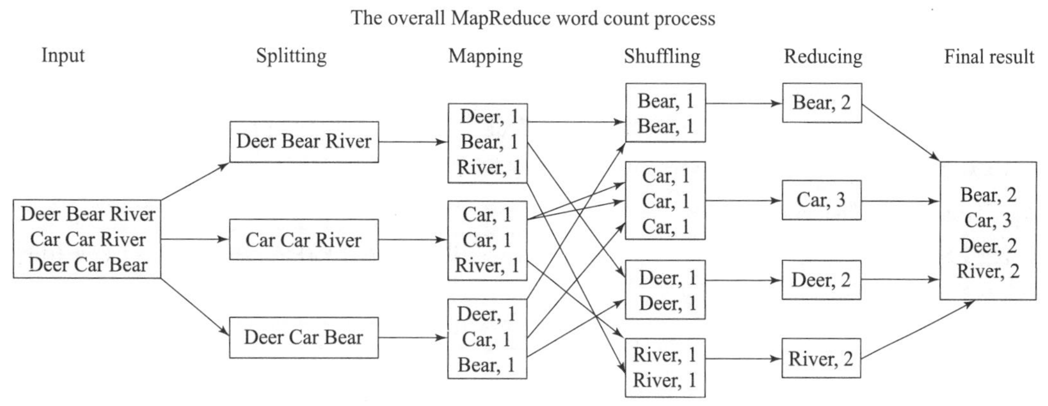

图 3 - 8　MapReduce 执行流程

通过图 3 - 8，可以看出，该计算过程大致分为 Input、Splitting（切片）、Mapping、Shuffling、Reducing、Final result 六个阶段，每个阶段完成各自的工作。

1. Splitting 阶段

该阶段，是将大文件拆分成若干个小文件。这一个过程由 MR 框架来完成。那么，框架是根据什么来将大文件拆分成小文件的呢？每个小文件是按照 max(min. split,min(max. split,block)) 来计算每个分片的大小的。其中，min. split 指的是允许分片的最小值，max. split 指的是允许分片的最大值，block 指的是数据块的大小。这样设计的好处是，拆分后的大小正好为一个 datanode 中数据块的大小，这样在计算时，就可以直接在 datanode 中计算，不需要再去其他 datanode 中取数据，减少了网络传输。通过图 3-8 可以看出，每个输入分片（Input Split）针对一个 Map 任务，将原始数据的每一行作为 Map 的输入。这就是 MapReduce 的分片策略。

2. Mapping 阶段

MapReduce 的思想就是“分而治之”，Mapper 负责“分”，即把复杂的任务分解为若干个“简单的任务”执行。其中，“简单的任务”指：数据或计算规模相对于原任务要缩减；就近计算，即会被分配到存放了所需数据的节点进行计算；这些小任务可以并行计算，彼此间几乎没有依赖关系。在 Mapping 阶段，将读取每一个数据片段，启动每一个 Map 任务，并行执行计算。

在 Mapping 阶段中，输入、输出均是以（Key,Value）的形式出现的。Hadoop 自带了好几个输入格式。其中有一个抽象类叫 fileinputformat，所有操作文件的 inputformat 类都是从它那里继承功能和属性。当开启 Hadoop 作业时，fileinputformat 会得到一个路径参数，这个路径内包含了所需要处理的文件，fileinputformat 会读取这个文件夹内的所有文件（注：默认不包括子文件夹内的），然后会把这些文件拆分成一个或多个 inputsplit。可以通过 jobconf 对象的 setinputformat( ) 方法来设定应用到自己的作业输入文件上的输入格式。默认的输入格式是 TextInputformat。Key 为 longwritable 类型的行的字节偏移量；Value 为该行的内容。

3. Shuffling 阶段

MapReduce 确保每个 Reducer 的输入都是按键排序的。系统执行排序的过程（即将 Map 输出作为 Reducer 的输入）称为 Shuffle。Shuffle 阶段比较复杂，可以分为排序（Sorting）和分组（Grouping）阶段。

经过 Shuffling 阶段后，将各个 Map 中的输出结果按照字典顺序排序，并且将 key 相同的划为一组，作为 Reducer 端的输入。

4. Reducing 阶段

对 Shuffling 阶段输出的结果进行统计分析，再将结果以 Key、Value 的形式输出。一般来说，如果任务量多，就要分配多个 Reduce，可以直接在代码中指定，如果没有指定，默认为 1。

例：一个简单的 MapReduce 操作

写一个 MapReduce 案例，这样有助于更好地理解 MapReduce 原理。案例需求：

①HDFS 目录/user/test/wc. txt 下有一个海量数据文件。

②写一个 MapReduce 代码，统计 wc. txt 文件中每个单词出现的次数。

根据需求和前面介绍的 MapReduce 运行原理，需要完成下列任务：

①准备测试数据 wc. txt，并上传到 HDFS 目录/user/test/中。

②编写 Mapper 类。

③编写 Reducer 类。

④编写 RunJob 类。

⑤把项目打包成 jar 文件。

准备测试数据

1. 准备测试数据

```
[root@YYR ~]# vi wc.txt
Hadoop is good
I like Hadoop
Hadoop includes Map and Reduce
We are all using Hadoop
Map and Reduce is very hard
```

上传到 HDFS 集群中：

```
[root@YYR ~]# hdfs dfs -put wc.txt /user/test/
```

2. 编写 Mapper 类

使用 Eclipse 构建一个 Java 项目，导入所有 Hadoop 的 jar 包，如图 3 -9 所示。代码如下：

```
package com.mr.wc;

import java.io.IOException;
import org.apache.hadoop.io.IntWritable;
import org.apache.hadoop.io.LongWritable;
import org.apache.hadoop.io.Text;
import org.apache.hadoop.mapreduce.Mapper;

/**
 * Map 任务执行的类
 * @author root
```

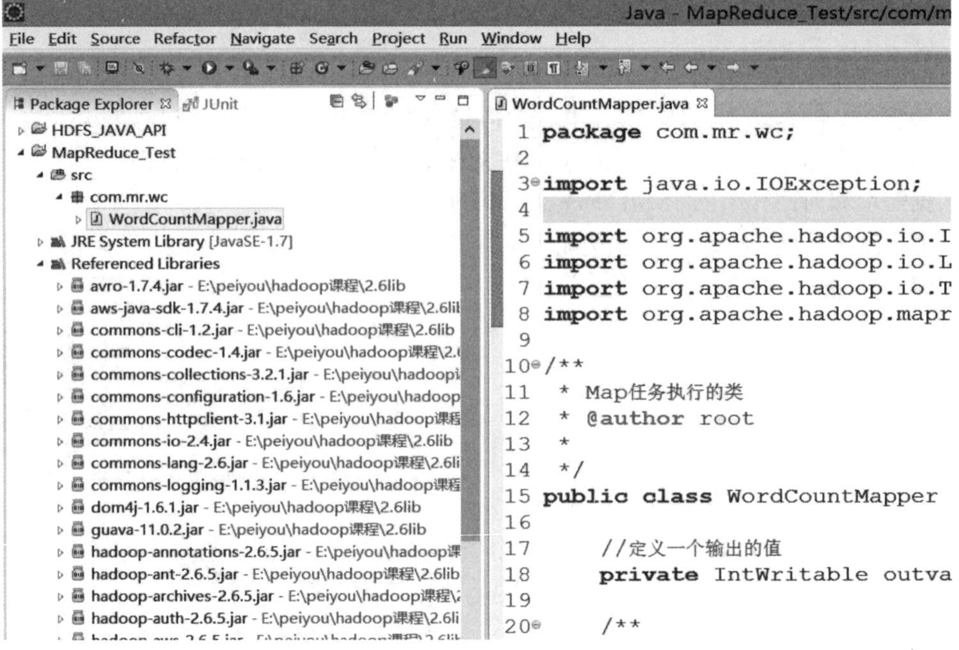

图 3－9　jar 包导入

```
*
*/
public class WordCountMapper extends Mapper < LongWritable, Text,Text,IntWritable >{

    //定义一个输出的值
    private IntWritable outvalue = new IntWritable(1);

    /**
    * Map 方法是一个循环调用的方法,读取分片(input split)每一行调用一次
     * @param key:代表改行的偏移量
     * @param value:代表改行内容
    */
    protected void map(LongWritable key,Text value,Context context)
            throws IOException,InterruptedException {
        String line = value.toString();
```

```
        String words[] = line.split(" ");
        for(String w:words) {
            Text outkey = new Text(w);
            //以单词为 Key,1 作为 Value 输出
            context.write(outkey,outvalue);
        }
    }
}
```

3. 编写 Reducer 类

```
package com.mr.wc;

import java.io.IOException;
import org.apache.hadoop.io.IntWritable;
import org.apache.hadoop.io.Text;
import org.apache.hadoop.mapreduce.Reducer;

/**
 * Reduce 任务执行的类
 * @author root
 *
 */
public class WordCountReduce extends Reducer < Text, IntWritable,Text,IntWritable >{

    /**
 *     Reduce 方法是一个循环调用的方法,一组(Group)数据调用一次
     * @param key:代表 Map 任务输出的 Key(在当前案例下,就是 word)
    * @param iter:因为一组数据包含有多条,所以通过迭代器来获取这一组的所有 Value
    */
     protected void reduce ( Text key, Iterable < IntWritable >iter,
             Context arg2 ) throws IOException,InterruptedException {

```

```
        int sum=0;
        for(IntWritable i:iter){//迭代器中的Value其实就是数字1
            sum=sum+i.get();//累加
        }
        /*由于在Shuffle过程中,按照单词(word)分组,所以这一组都是一个单词*/
    //把1累加之后得到单词的数量。
        arg2.write(key,new IntWritable(sum));
    }
}
```

4. 编写 RunJob 类

编写 RUNJOB 类

```
package com.mr.wc;

import java.io.IOException;
import org.apache.hadoop.conf.Configuration;
import org.apache.hadoop.fs.FileSystem;
import org.apache.hadoop.fs.Path;
import org.apache.hadoop.io.IntWritable;
import org.apache.hadoop.io.Text;
import org.apache.hadoop.mapreduce.Job;
import org.apache.hadoop.mapreduce.lib.input.FileInputFormat;
import org.apache.hadoop.mapreduce.lib.output.FileOutputFormat;

/**
 * 执行Job的入口类
 * @author root
 *
 */
public class RunJob {

    public static void main(String[]args){
        //本地测试运行需要
//    System.setProperty("hadoop.home.dir","E:\\peiyou\\hadoop-2.6.5");
```

```
Configuration config = new Configuration();
/* 如果需要在 Windows 系统下测试运行,不能加载服务器的 yarn - site.xml * /
try {
    FileSystem fs = FileSystem.get(config);
    //创建一个 Job
    Job job = Job.getInstance(config);
    job.setJobName("wordcount");
    job.setJarByClass(RunJob.class);

    job.setMapperClass(WordCountMapper.class);
    job.setReducerClass(WordCountReduce.class);

    job.setMapOutputKeyClass(Text.class);
    job.setMapOutputValueClass(IntWritable.class);

    job.setCombinerClass(WordCountReduce.class);
    //设置当前 Job 中的 Reduce 任务数量为 1
    job.setNumReduceTasks(1);
    //设置当前 Job 的输入数据为 /user/test/wc.txt
    FileInputFormat.addInputPath(job,new Path("/user/test/wc.txt"));
    //设置当前 Job 的输出目录为 /user/test/output/
    Path outPath = new Path("/user/test/output/");
    if(fs.exists(outPath)) {//如果输出目录存在则删除
    fs.delete(outPath,true);
    }
    FileOutputFormat.setOutputPath(job,outPath);

    //正式提交 Job 到服务器运行
    boolean stauts = job.waitForCompletion(true);

    if(stauts) {//如果返回 true,则运行成功
        System.out.println("success");
    }
} catch (IOException | ClassNotFoundException | InterruptedException e) {
```

```
            e.printStackTrace();
        }

    }
}
```

5. 把项目打包成 jar 文件

把项目打包成 JAR 文件

第一步：右击项目名字，选择“Export”，如图 3－10 所示。

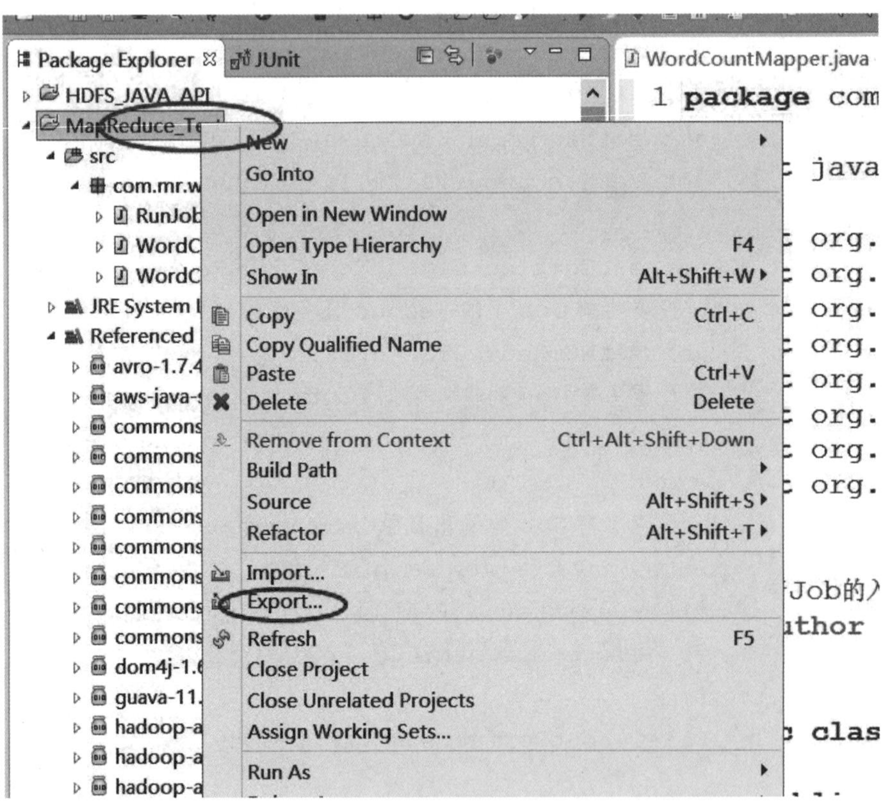

图 3－10　选择输出项

第二步：选择 jar 文件，单击“Next”按钮，如图 3－11 所示。

第三步：选择文件路径，单击“Finish”按钮，如图 3－12 所示。

完成上述操作后，就已经写好第一个 MapReduce 程序了。数据清洗功能的代码开发和前面的 WordCount 案例一样，需要编写三个核心代码。完成代码后，先不执行。因为 MapReduce 程序的运行需要 YARN 分布式资源管理系统。

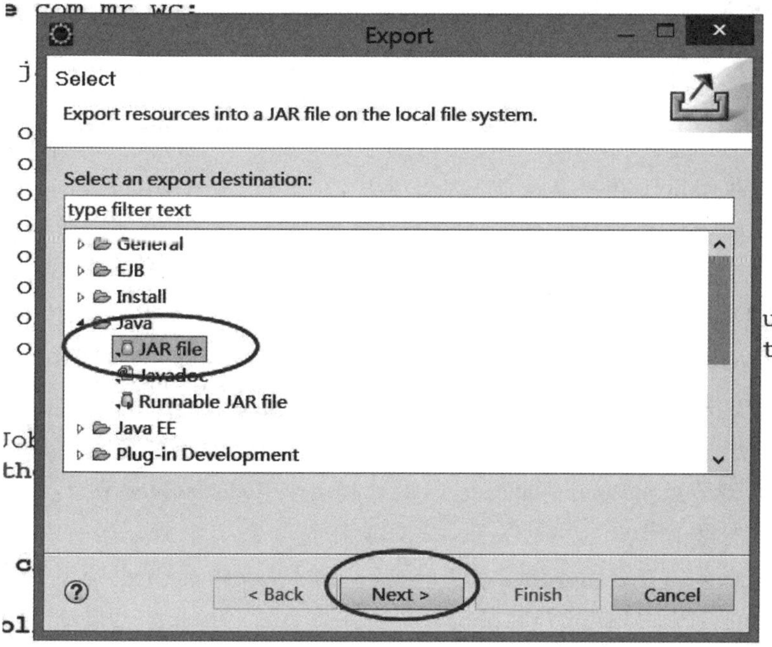

图 3－11　选择 jar 文件

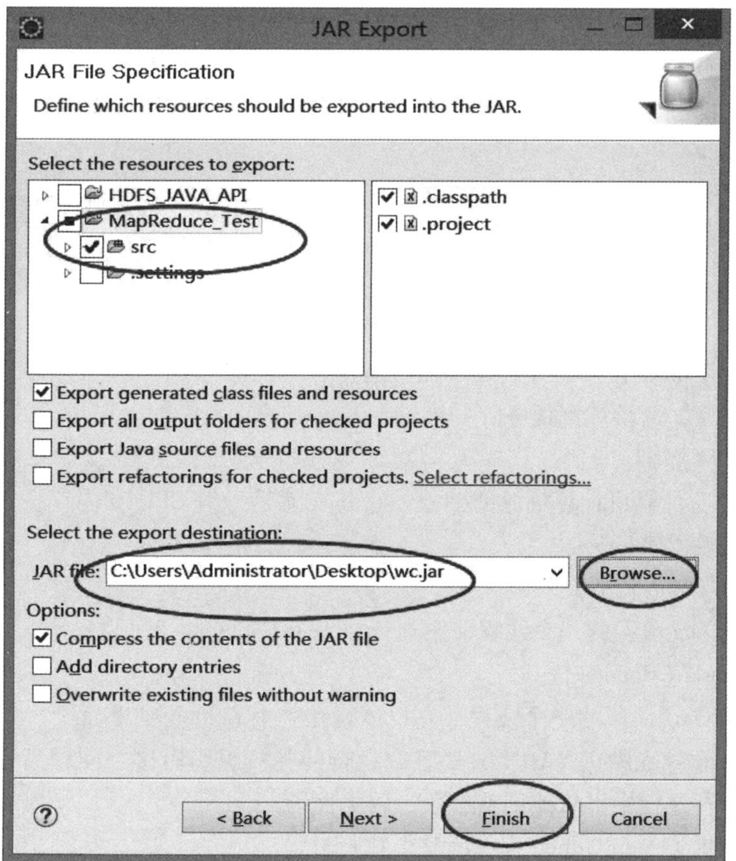

图 3－12　完成 jar 文件输出

任务3　YARN 分布式资源管理平台的资源调度

YARN 是数据清洗 MapReduce 程序运行的平台，是一个分布式资源管理系统，它是从 MRv1 发展过来的。MRv1 在 Hadoop 1. x 版本中广泛使用。本书采用 Hadoop 2. x，所以不对 MRv1 做过多描述。

YARN 是由早期的 MapReduce 经过架构变换而来的，主要思路是把原有 MapReduce 架构分为资源管理和监控调度两部分。YARN 舍弃了 MRv1 中的 JobTrack 和 TaskTrack，采用一种新的 MRAppYYR 进行管理，并与 ResourceManager 和 NodeManger 两个守护进程一起协同调度和控制任务，避免单一进程服务的管理和调度负载过重。

YARN 的原理是把 MRv1 的 JobTracker 分成两部分，即资源管理和工作任务调度。有一个全局的资源管理节点 ResourceManager（RM）和多个 NodeManager 节点。ResourceManager 和 NodeManager 的每个节点是主从关系，YARN 在整个系统中调配所有应用的资源。ResourceManager 接收到客户端任务请求后，会交给某个 NodeManager 并启动一个进程 MRAppYYR 负责任务的完成。MRAppYYR 分配任务到其他 DataNode 节点并启动 TaskYYR 进程。

YARN 中的核心概念如下：

- ResourceManager（RM）
  - 是在整个框架系统中负责仲裁应用程序之间资源的最终决定者；负责协调集群上计算资源的分配；调度、启动每一个 Job 所属的 ApplicationYYR；监控 ApplicationYYR 的存在情况。
- NodeManager（NM）
  - 功能比较专一，根据要求启动和监视集群中机器的计算容器 Container。负责 Container 状态的维护，与 RM 保持联系，并汇报该节点资源使用情况。
- ApplicationMaster（AM）
  - 负责一个 Job 生命周期内的所有工作。注意，每一个 Job 都有一个 ApplicationYYR。它和 MapReduce 任务一样，在容器中运行。AM 通过与 RM 交互获取资源，然后通过与 NM 交互，启动计算任务。
- Container（容器）
  - 是 YARN 对计算机计算资源的抽象，它其实就是一组 CPU 和内存资源，所有的应用都会运行在 Container 中。
- MapTask（Map 任务）
  - 执行 Mapping 阶段的一个线程，它运行在一个 Container 里面。
- ReduceTask（Reduce 任务）
  - 执行 Reducing 阶段的一个线程，它也运行在一个 Container 里面。

那么 MapReduce 程序在 YARN 中是怎么运行的呢？想明白这个问题后，才能在以后的数据清洗程序执行过程中不断地调整参数、调优代码。

图 3 – 13 展示了 MapReduce 程序的整个执行过程。

步骤 1：客户端程序向 ResourceManager 提交应用，并请求一个 ApplicationYYR 实例。一

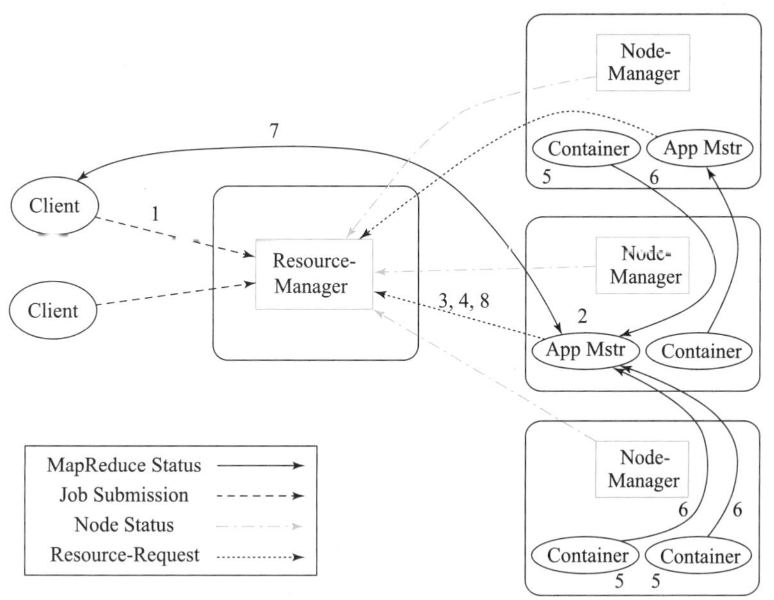

图 3－13　MapReduce 程序的执行过程

个 Job 对应一个 ApplicationYYR。

步骤 2：ResourceManager 找到可以运行一个 Container 的 NodeManager，并在这个 Container 中启动 ApplicationYYR 实例。

步骤 3：ApplicationYYR 向 ResourceManager 进行注册，注册之后，客户端就可以查询 ResourceManager，从而获得自己 ApplicationYYR 的详细信息，以后就可以和自己的 ApplicationYYR 直接交互了。

步骤 4：一般情况下，ApplicationYYR 根据 Resource－Request（资源请求）协议向 ResourceManager 发送 Resource－Request。ResourceManager 发现集群的某些符合条件的 NodeManager 中还有资源，则分配 Container（资源）给 ApplicationYYR。

步骤 5：当 Container 被成功分配之后，ApplicationYYR 通过向 NodeManager 发送 Container－Launch－Specification 信息来启动 Container，Container－Launch－Specification 信息包含了能够让 Container 和 ApplicationYYR 交流所需要的资料。

步骤 6：应用程序的代码在启动的 Container 中运行，并把运行的进度、状态等信息通过 Application－Specific 协议发送给 ApplicationYYR。

步骤 7：在应用程序运行期间，提交应用的客户端主动和 ApplicationYYR 交流，获得应用的运行状态、进度更新等信息，交流的协议也是 Application－Specific 协议。

步骤 8：一旦应用程序执行完成并且所有相关工作也已经完成，则 ApplicationYYR 向 ResourceManager 取消注册并关闭，用到的所有的 Container 也归还给系统。

例：在 YARN 中运行 MapReduce 程序

现在可以提交之前写好的 WordCount 程序到 YARN 中。同时，也可以验证上一任务中的应用程序执行步骤。假设每个人的 Hadoop 环境都已经部署成功，并且全部启动成功。实现下面四个任务，完成 MapReduce 程序的执行过程：

①上传程序的 jar 包到 Linux 主机（YYR 机器）的/home/目录下。

②执行命令，提交 jar 程序到 YARN 集群。

③在 Web 界面上查看运行状态。

④在 HDFS 中查看程序运行的结果。

在 YARN 中运行 MAPREDUCE 程序

1. 上传 wc. jar 包到 YYR 机器（图 3-14）

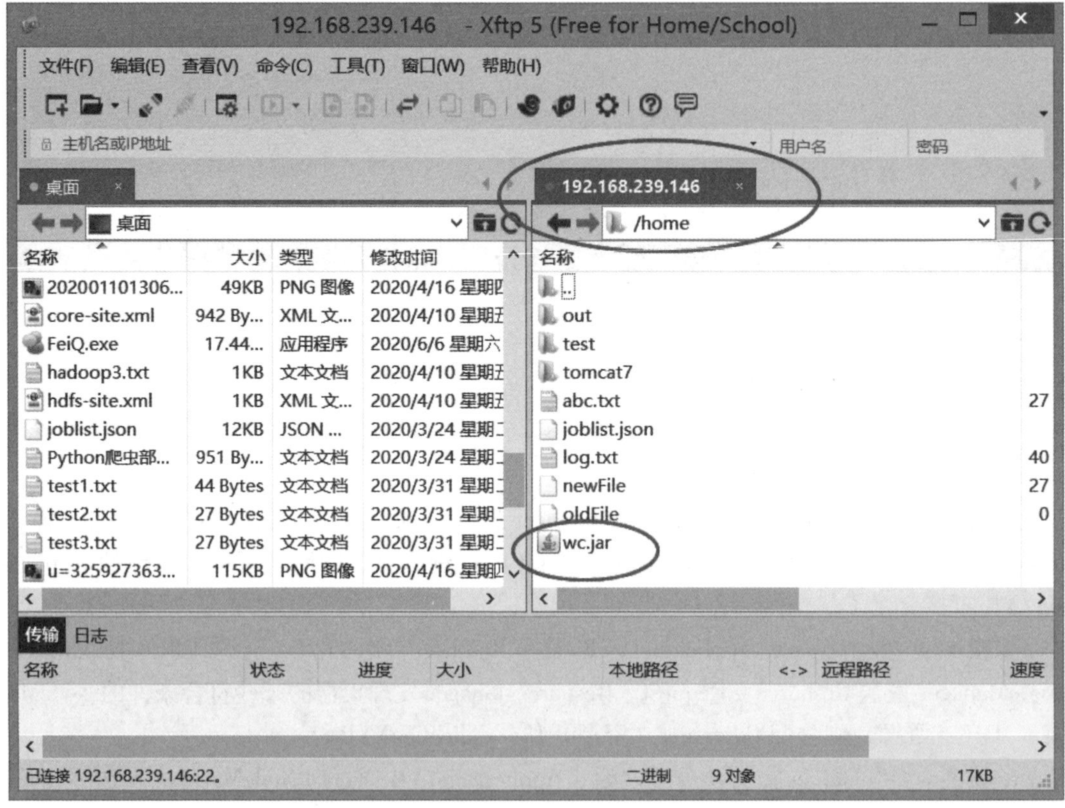

图 3-14　上传 wc. jar 包到 YYR 机器

2. 执行命令，提交程序

```
[root@YYR home]# ls -l
总用量 48
-rw-r--r--. 1 root root    27 4 月  1 01:48 abc.txt
-rw-r--r--. 1 root root 12487 3 月 28 23:44 joblist.json
-rw-r--r--. 1 root root    40 4 月  1 02:03 log.txt
-rw-r--r--. 1 root root    27 4 月 17 01:16 newFile
-rw-r--r--. 1 root root     0 4 月 17 01:16 oldFile
```

```
drwxr-xr-x. 2 root root  4096 4 月   1 02:03 out
drwxr-xr-x. 2 root root  4096 4 月   1 01:54 test
drwxr-xr-x. 9 root root  4096 4 月   1 03:06 tomcat7
-rw-r--r--. 1 root root  4830 4 月  17 06:12 wc.jar
[root@YYR home]# hadoop jar /home/wc.jar com.mr.wc.RunJob
```

3. 在 Web 界面上查看程序运行状态（图 3－15）

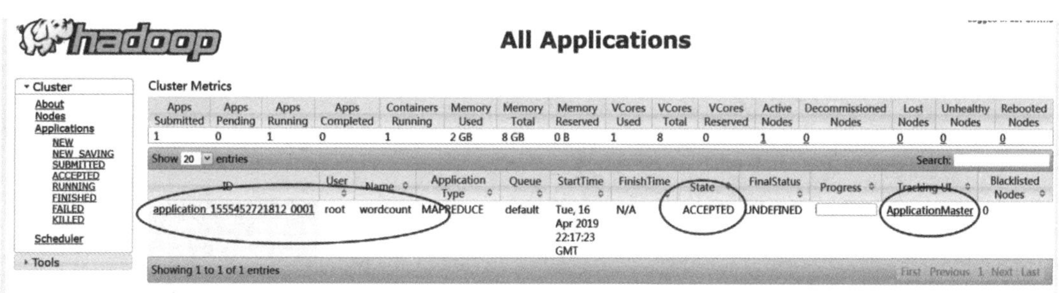

图 3－15　在 Web 界面上查看程序运行状态

立即刷新页面，状态变成“RUNNING”，单击“ApplicationMaster”，如图 3－16 所示。

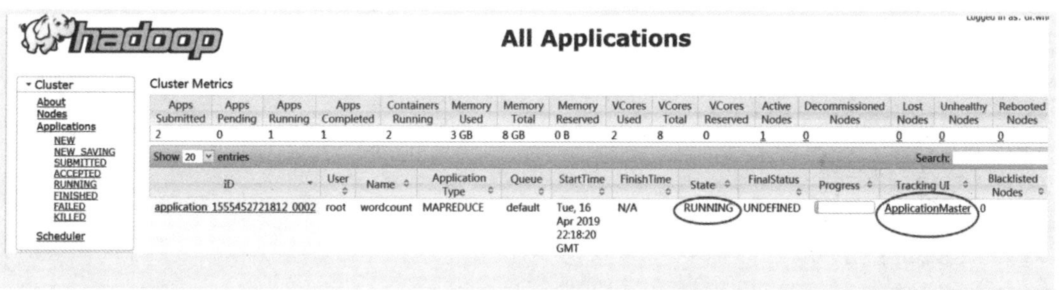

图 3－16　在 RUNNING 状态下查看

可以看到每个任务执行进度，如图 3－17 所示。

图 3－17　查看每个任务进度

回到首页之后，可以看到工作全部执行完成，如图 3－18 所示。

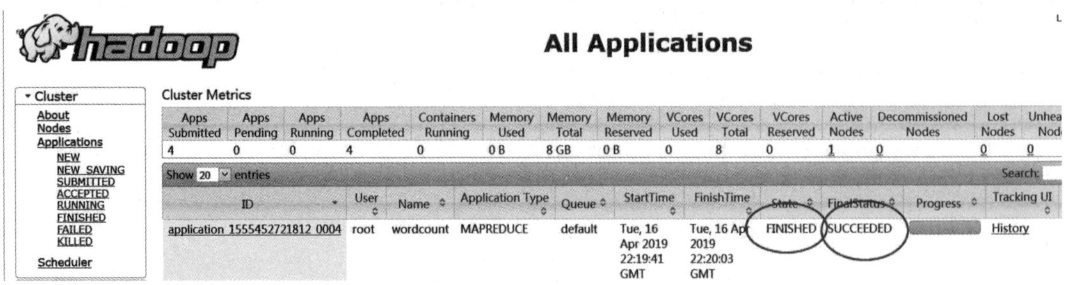

图 3－18　工作全部执行完成

4. 在 HDFS 上查看运行结果

```
[root@YYR home]# hadoop jar /home/wc.jar com.mr.wc.RunJob
20/04/17 06:19:38 WARN util.NativeCodeLoader:Unable to load native-hadoop library for your platform... using builtin-java classes where applicable
20/04/17 06:19:39 INFO client.RMProxy:Connecting to ResourceManager at guangzhou001/192.168.239.146:8032
20/04/17 06:19:40 WARN mapreduce.JobResourceUploader: Hadoop command-line option parsing not performed. Implement the Tool interface and execute your application with ToolRunner to remedy this.
20/04/17 06:19:40 INFO input.FileInputFormat:Total input paths to process:1
20/04/17 06:19:40 INFO mapreduce.JobSubmitter: number of splits:1
20/04/17 06:19:40 INFO mapreduce.JobSubmitter:Submitting tokens for job:job_1555452721812_0004
20/04/17 06:19:41 INFO impl.YarnClientImpl:Submitted application application_1555452721812_0004
20/04/17 06:19:41 INFO mapreduce.Job:The url to track the job:http://guangzhou001:8088/proxy/application_1555452721812_0004/
20/04/17 06:19:41 INFO mapreduce.Job: Running job: job_1555452721812_0004
20/04/17 06:19:48 INFO mapreduce.Job:Job job_1555452721812_0004 running in uber mode:false
20/04/17 06:19:48 INFO mapreduce.Job:map 0% reduce 0%
20/04/17 06:19:54 INFO mapreduce.Job:map 100% reduce 0%
```

```
20/04/17 06:20:04 INFO mapreduce.Job:map 100% reduce 100%
20/04/17 06:20:04 INFO mapreduce.Job:Job job_1555452721812_0004 completed successfully
20/04/17 06:20:04 INFO mapreduce.Job:Counters:49
    File System Counters
        FILE:Number of bytes read =169
        FILE:Number of bytes written =214479
        FILE:Number of read operations =0
        FILE:Number of large read operations =0
        FILE:Number of write operations =0
        HDFS:Number of bytes read =218
        HDFS:Number of bytes written =103
        HDFS:Number of read operations =6
        HDFS:Number of large read operations =0
        HDFS:Number of write operations =2
    Job Counters
        Launched map tasks =1
        Launched reduce tasks =1
        Data - local map tasks =1
        Total time spent by all maps in occupied slots (ms) =4173
        Total time spent by all reduces in occupied slots (ms) =6722
        Total time spent by all map tasks (ms) =4173
        Total time spent by all reduce tasks (ms) =6722
        Total vcore - milliseconds taken by all map tasks =4173
        Total vcore - milliseconds taken by all reduce tasks =6722
        Total megabyte - milliseconds taken by all map tasks =4273152
        Total megabyte - milliseconds taken by all reduce tasks =6883328
    Map - Reduce Framework
        Map input records =5
        Map output records =22
        Map output bytes =200
        Map output materialized bytes =169
        Input split bytes =106
```

```
        Combine input records =22
        Combine output records =15
        Reduce input groups =15
        Reduce shuffle bytes =169
        Reduce input records =15
        Reduce output records =15
        Spilled Records =30
        Shuffled Maps =1
        Failed Shuffles =0
        Merged Map outputs =1
        GC time elapsed (ms) =95
        CPU time spent (ms) =1560
        Physical memory (bytes) snapshot =333647872
        Virtual memory (bytes) snapshot =1682927616
        Total committed heap usage (bytes) =164040704
    Shuffle Errors
        BAD_ID =0
        CONNECTION =0
        IO_ERROR =0
        WRONG_LENGTH =0
        WRONG_MAP =0
        WRONG_REDUCE =0
    File Input Format Counters
        Bytes Read =112
    File Output Format Counters
        Bytes Written =103
Success
[root@YYR home]# hdfs dfs -cat /user/test/output/*
20/04/17 06:30:10 WARN util.NativeCodeLoader:Unable to load native-hadoop library for your platform... using builtin-java classes where applicable
Hadoop4
I  1
Map2
Reduce2
We1
all1
```

```
and2
are1
good1
hard1
includes1
is2
like1
using1
very1
```

至此，整个 MapReduce 程序执行成功。以后的数据清洗程序也会提交给 YARN 运行。需要提醒的是，要根据 Web 界面的信息来查看程序运行情况。

学习笔记

补充知识——HOST 选择算法

补充知识——文件切分算法

项目四
数据仓库——Hive的搭建与应用

项目描述

假设需要完成一个大数据离线分析项目，从现在开始进行 Hadoop 离线分析平台的学习。为什么在数据分析阶段要使用 Hive 来做平台支撑呢？

在安装好的 Hadoop 集群中，运行了 MapReduce 任务，编写 MapReduce 程序须具备 Java 基础知识，但是大数据应用场景中，不一定都是由开发人员来完成数据分析工作，某些专职人员会根据公司的历史数据或实时采集的数据进行分析，从而产生对公司运营方面的决策。在此情况下，没必要让他们懂得一门编程语言。SQL 语言的学习成本低，并且简单易懂。大部分的数据分析人员具备 SQL 语言的编写能力，于是越来越多的项目团队引入 Hive 来帮助专职数据分析师完成基于大数据场景下的数据分析。

本项目中要掌握 Hive 搭建与应用。

项目分析

Hive 用来建立企业级数据仓库，将企业不同的数据源上的数据按照一定的标准统一存储在 Hive 中，使用 MapReduce 分布式计算框架进行计算，并得到结果。

数据仓库一般从用户实际需求出发，将不同平台的数据源按设定主题进行划分、整合，具有较高的抽象性。面向主题的数据组织方式，是在较高层次对分析对象数据的一个完整、统一并一致的描述，能完整及统一地刻画各个分析对象所涉及的有关企业的各项数据，以及数据之间的联系。

通过以下几个任务来完成对知识点 Hive 的学习：

任务 1：Hive 搭建的准备工作。

任务 2：Hive 组件的新增工作。

任务 3：用 Hive 来存储数据并执行查询分析。

数据仓库

任务1　Hive 搭建的准备工作 

HIVE 的由来

1. Hive 搭建背景知识

Hive 由 Facebook 实现并开源，是一个基于 Hadoop 的数据仓库工具，能够将结构化的数据映射为一张数据库表，并提供 HQL（Hive SQL）查询功能。底层数据存储在 HDFS 上。Hive 的本质是将 SQL 语句转换为 MapReduce 任务运行，使不熟悉 MapReduce 的用户很方便地利用 HQL 处理和计算 HDFS 上的结构化的数据，对于离线的批量数据计算比较适合。

数据仓库是一个面向主题的、集成的、相对稳定的、反映历史变化的数据集合，用于支持管理决策。这一概念是由数据仓库之父比尔·恩门（Bill Inmon）在 1991 年出版的“Building the Data Warehouse”（《建立数据仓库》）一书中提出的，并得到广泛的推广应用。

Hive 依赖于 HDFS 来对数据进行存储，它将 HQL 转换成 MapReduce 执行，所以说 Hive 是基于 Hadoop 的一个数据仓库工具，实质就是一款基于 HDFS 的 MapReduce 计算框架，对存储在 HDFS 中的数据进行分析和管理。图 4－1 所示是 Hive 的官方图解。

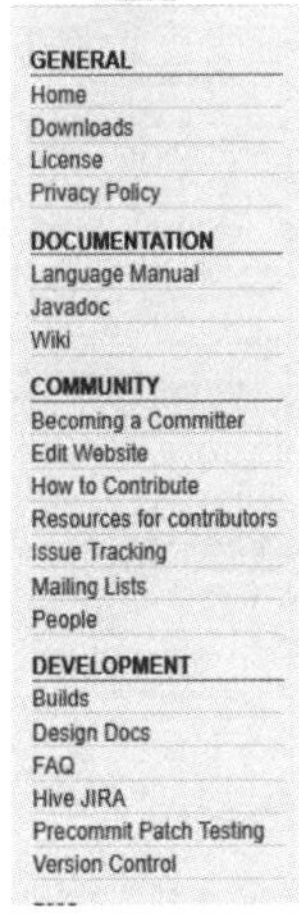

图 4－1　Hive 官网显示

通过学习，发现 Hive 的安装是以 Hive 元数据的存储与管理为原则进行部署分类的，主要有以下三种模式。

■ 单机模式

Hive 的安装包中包含了一个内存数据库 Derby，可对其进行直接安装配置，但实际的生产环境中，几乎没有企业选择采用这样的部署方式，因为元数据不利于存储和管理，因此，对这种模式了解即可，不需要进行实际操作。

■ 远程数据库模式

在该安装部署模式下，Hive 将不再使用自带的内存数据库 Derby，而选择通过网络的方式访问部署在另一台服务器上的 MySQL 或者其他关系型数据库。在这种模式下，通常需要准备两台服务器，一台用来安装 MySQL，一台用来安装 Hive。这种部署模式一般在开发和测试阶段使用较多。

■ 远程数据库服务模式

该部署模式是在远程数据库模式的方式上做了一层解耦。在远程数据库模式中，Hive 需要直接连接到 MySQL 数据库，这样耦合性比较高，因此，可以选择第三台服务器，专门用来与数据库进行交互，而 Hive 只需要与该服务器的元数据服务交互即可，这样 Hive 不需要直接连接到 MySQL 数据库，当需要更换 Hive 的元数据存储方式时，Hive 不会受到任何影响，并且元数据对外服务，可以让多个客户端同时连接。该模式下，需要准备三台服务器，一台安装 MySQL，一台作为元数据服务节点，一台只需要连接元数据服务。这种方式在生产环境中使用比较多。

2. Hive 版本选择

通过学习，了解到 Hive 依赖于 HDFS 来对数据进行存储，类 SQL 的执行底层是 MapReduce 任务，因此 Hive 版本的选择要依赖于 Hadoop。通过官网截图（图 4 – 2），可以比较 Hive 与 Hadoop 版本之间的依赖关系。

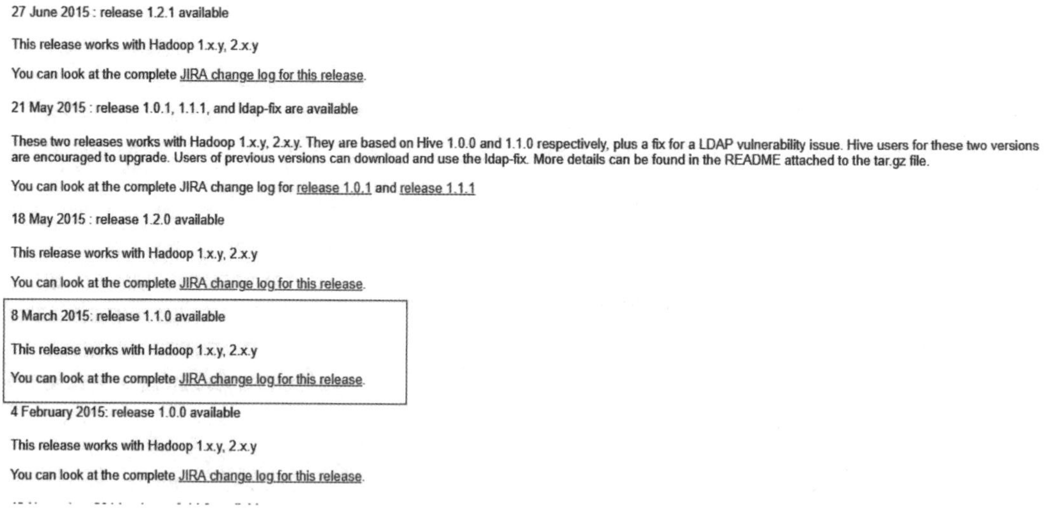

27 June 2015 : release 1.2.1 available

This release works with Hadoop 1.x.y, 2.x.y

You can look at the complete JIRA change log for this release.

21 May 2015 : release 1.0.1, 1.1.1, and ldap-fix are available

These two releases works with Hadoop 1.x.y, 2.x.y. They are based on Hive 1.0.0 and 1.1.0 respectively, plus a fix for a LDAP vulnerability issue. Hive users for these two versions are encouraged to upgrade. Users of previous versions can download and use the ldap-fix. More details can be found in the README attached to the tar.gz file.

You can look at the complete JIRA change log for release 1.0.1 and release 1.1.1

18 May 2015 : release 1.2.0 available

This release works with Hadoop 1.x.y, 2.x.y

You can look at the complete JIRA change log for this release.

8 March 2015: release 1.1.0 available

This release works with Hadoop 1.x.y, 2.x.y

You can look at the complete JIRA change log for this release.

4 February 2015: release 1.0.0 available

This release works with Hadoop 1.x.y, 2.x.y

You can look at the complete JIRA change log for this release.

图 4 – 2　Hive 与 Hadoop 版本之间的依赖关系

官网指出，Hive 1. 1. 0 搭配的 Hadoop 版本是 1. x 或者 2. x，但是由于 1. x 中存在问题，所以一般会在 2. x 上搭建。

有些同学可能会提出疑问，Hive 2. x 也已经有了，并且适用于 Hadoop 2. x，那么为什么不使用 Hive 2. x？原因其实非常简单，在执行 MapReduce 任务时，发现即使一个简单的 WordCount 也需要三四十秒的时间，而执行 Hive 的 SQL 语句大概也需要这么久，所以 Hive 在考虑替换 Hive SQL 的执行引擎。如果搭建 Hive 2. x，在启动之后会看到图 4 – 3 中框起来的这句话。

```
SLF4J: Class path contains multiple SLF4J bindings.
SLF4J: Found binding in [jar:file:/opt/sxt/hive-2.3.4/lib/log4j-slf4j-impl-2.6.2.jar!/org/slf4j/impl/StaticLoggerBinder.class]
SLF4J: Found binding in [jar:file:/opt/sxt/hadoop-2.6.5/share/hadoop/common/lib/slf4j-log4j12-1.7.5.jar!/org/slf4j/impl/StaticLoggerBinder.class]
SLF4J: See http://www.slf4j.org/codes.html#multiple_bindings for an explanation.
SLF4J: Actual binding is of type [org.apache.logging.slf4j.Log4jLoggerFactory]

Logging initialized using configuration in jar:file:/opt/sxt/hive-2.3.4/lib/hive-common-2.3.4.jar!/hive-log4j2.properties
 Async: true
Hive-on-MR is deprecated in Hive 2 and may not be available in the future versions. Consider using a different execution engine (i.e. spark, tez) or using Hive 1.X releases.
hive>
```

图 4-3 Hive 2. x 上的搭建效果

框起来的内容表示：Hive 在 MapReduce 上的执行已经被弃用，在未来的版本中可能不能执行，推荐使用一个不同的执行引擎（例如 Spark 或者 Tez），或者使用 Hive 1. x 版本，因此，在之后的学习中使用的都是 1. 1. 0 的版本。

3. Hive 集群规划

Hive 集群规划见表 4-1。

表 4-1 Hive 集群规划

角色	YYR	Superintendent1	Superintendent2
MySQL	●	×	×
Metastore	×	●	×
CLI	×	×	●

■ MySQL

MySQL 是传统的数据库，具有开源免费的特点，在中小型企业中广泛应用。Hive 的搭建过程中，需要使用 MySQL 来存储元数据。Hive 的表中存储的是元数据，包括表的字段名称、表的字段类型、分区信息、权限管理等信息。那么，Hive 为什么需要使用 MySQL 作为其元数据的存储介质呢？这是因为在 Hive 中默认使用内存数据库 derby 作为存储介质，此时在哪一级目录启动，Hive 就会在此目录下生成一个 metastore_db 文件，这种情况下元数据不利于存储和管理，所以需要集中对其进行管理，于是便将元数据写入 MySQL 中进行统一管理。

■ Metastore

Metastore 是 Hive 启动运行时必不可少的组件，Hive 的元数据存储在 MySQL 中，但需要对外提供服务，因此要将对 MySQL 的连接属性等封装起来，建立 MySQL 的连接，方便客户端进行 Hive 操作。Hive 的某些操作必然会触发元数据的修改，例如创建表、创建数据库等，因此，Metastore 要作为一个独立的服务运行，提供对 MySQL 中数据修改的连接。

■ CLI

CLI 即 Command Line Interface（命令行界面），用户可以通过 CLI 对 Hive 进行操作，可以提交 SQL 语句，进行复杂的查询分析。

前面对 Hive 中的主要角色做了介绍，这里选择远程数据库服务模式进行部署，使用三台服务器，分别部署安装 MySQL、CLI、Metastore。

4. 安装 MySQL

通过学习可知，要安装 Hive，需将 MySQL 作为其元数据的存储介质，则需要先在 Linux 上安装 MySQL。

（1）在 Linux 上使用 YUM 的方式安装 MySQL

使用 root 登录 YYR，输入以下命令安装 MySQL：

在 LINUX 上使用 YUM 的方式安装 MYSQL

```
# 查看本机是否安装了 MySQL
[root@YYR ~]# rpm -qa |grep mysql

# 安装 MySQL 关联组件
[root@YYR ~]# yum -y install mysql
已加载插件:fastestmirror
Loading mirror speeds from cached hostfile
 * base:mirrors.aliyun.com
 * extras:mirrors.aliyun.com
 * updates:mirrors.huaweicloud.com
软件包 1:mariadb-5.5.60-1.el7_5.x86_64 已安装并且是最新版本
无须任何处理

# 卸载 MySQL 无关组件
[root@YYR ~]# yum -y remove mysql
已加载插件:fastestmirror
正在解决依赖关系
--> 正在检查事务
---> 软件包 mariadb.x86_64.1.5.5.60-1.el7_5 将被删除
--> 解决依赖关系完成

依赖关系解决 …(省略处理过程)
删除:
  mariadb.x86_64 1:5.5.60-1.el7_5

完毕!

# 安装 wget 组件,如果本机已经安装,请忽略本步骤
[root@YYR ~]# yum -y install wget
已加载插件:fastestmirror
```

```
Loading mirror speeds from cached hostfile
 * base:mirrors.aliyun.com
 * extras:mirrors.aliyun.com
 * updates:mirrors.huaweicloud.com
正在解决依赖关系
-->正在检查事务
--->软件包 wget.x86_64.0.1.14-18.el7 将被安装
-->解决依赖关系完成

依赖关系解决
 ... (省略处理过程)
[root@YYR ~]# wget http://repo.mysql.com/mysql57-community-release-el7-10.noarch.rpm
2020-05-07 09:49:15 (11.5 KB/s) - 已保存"mysql57-community-release-el7-10.noarch.rpm"[25548/25548])

# 安装 MySQL 5.7 rpm 包
[root@YYR ~]# rpm -Uvh mysql57-community-release-el7-10.noarch.rpm
警告:mysql57-community-release-el7-10.noarch.rpm:头 V3 DSA/SHA1 Signature,密钥 ID 5072e1f5:NOKEY
准备中...                    ###################################[100%]
正在升级/安装...
1:mysql57-community-release-el7-0
                             ###################################[100%]
[root@YYR ~]# yum listrepo |grep mysql
没有该命令:listrepo。请使用/usr/bin/yum --help

# 查看可安装的 MySQL 组件
[root@YYR ~]# yum repolist |grep mysql
mysql-connectors-community/x86_64   MySQL Connectors Community   108
mysql-tools-community/x86_64        MySQL Tools Community        89
mysql57-community/x86_64            MySQL 5.7 Community Server   347
```

```
# 安装 MySQL Server 5.7
[root@YYR ~]# yum install -y mysql - community - server
已加载插件:fastestmirror
Loading mirror speeds from cached hostfile
 * base:mirrors.aliyun.com
 * extras:mirrors.aliyun.com
 * updates:mirrors.huaweicloud.com
正在解决依赖关系
--> 正在检查事务
---> 软件包 mysql - community - server.x86_64.0.5.7.26 -1.el7 将被安装
--> 正在处理依赖关系 mysql - community - common(x86 -64) =5.7.26 -1.el7,它被软件包 mysql - community - server -5.7.26 -1.el7.x86_64 需要
--> 正在处理依赖关系 mysql - community - client(x86 -64) > =5.7.9,它被软件包 mysql - community - server -5.7.26 -1.el7.x86_64 需要
--> 正在处理依赖关系 net - tools,它被软件包 mysql - community - server -5.7.26 -1.el7.x86_64 需要
--> 正在检查事务
---> 软件包 mysql - community - client.x86_64.0.5.7.26 -1.el7 将被安装
--> 正在处理依赖关系 mysql - community - libs(x86 -64) > =5.7.9,它被软件包 mysql - community - client -5.7.26 -1.el7.x86_64 需要
---> 软件包 mysql - community - common.x86_64.0.5.7.26 -1.el7 将被安装
---> 软件包 net - tools.x86_64.0.2.0 -0.24.20131004git.el7 将被安装
--> 正在检查事务
---> 软件包 mariadb - libs.x86_64.1.5.5.60 -1.el7_5 将被取代
--> 正在处理依赖关系 libmysqlclient.so.18()(64bit),它被软件包 2:postfix -2.10.1 -6.el7.x86_64 需要
--> 正在处理依赖关系 libmysqlclient.so.18(libmysqlclient_18)(64bit),它被软件包 2:postfix -2.10.1 -6.el7.x86_64 需要
---> 软件包 mysql - community - libs.x86_64.0.5.7.26 -1.el7 将被舍弃
--> 正在检查事务
```

```
---> 软件包 mysql - community - libs - compat.x86_64.0.5.7.26 -
  1.el7 将被舍弃
---> 软件包 postfix.x86_64.2.2.10.1 -6.el7 将被升级
---> 软件包 postfix.x86_64.2.2.10.1 -7.el7 将被更新
--> 解决依赖关系完成
依赖关系解决
   ... (省略处理过程)
已安装:
  mysql - community - libs.x86_64 0:5.7.26 -1.el7
mysql - community - libs - compat.x86_64 0:5.7.26 -1.el7
       mysql - community - server.x86_64 0:5.7.26 -1.el7
作为依赖被安装:
  mysql - community - client.x86_64 0:5.7.26 -1.el7
   mysql - community - common.x86_64 0:5.7.26 -1.el7
   net - tools.x86_64 0:2.0 -0.24.20131004git.el7
作为依赖被升级:
  postfix.x86_64 2:2.10.1 -7.el7
替代:
  mariadb - libs.x86_64 1:5.5.60 -1.el7_5
完毕!
[root@YYR ~]#
```

启动 MYSQL 服务

(2) 启动 MySQL 服务

```
# 查看 MySQL 服务状态
[root@YYR ~]# service mysqld status
Redirecting to /bin/systemctl status mysqld.service
mysqld.service - MySQL Server
   Loaded:loaded (/usr/lib/systemd/system/mysqld.service;ena-
   bled;vendor preset:disabled)
   Active:inactive (dead)
     Docs:man:mysqld(8)
```

```
          http://dev.mysql.com/doc/refman/en/using-systemd.html

#启动 MySQL 服务
[root@YYR ~]# service mysqld start
Redirecting to /bin/systemctl start mysqld.service

#查看服务启动是否正常
[root@YYR ~]# service mysqld status
Redirecting to /bin/systemctl status mysqld.service
mysqld.service - MySQL Server
   Loaded: loaded (/usr/lib/systemd/system/mysqld.service; enabled; vendor preset: disabled)
   Active: active (running) since 二 2020-05-07 10:11:38 CST; 9s ago
     Docs: man:mysqld(8)
           http://dev.mysql.com/doc/refman/en/using-systemd.html
  Process: 13005 ExecStart=/usr/sbin/mysqld --daemonize --pid-file=/var/run/mysqld/mysqld.pid $MYSQLD_OPTS (code=exited, status=0/SUCCESS)
  Process: 12931 ExecStartPre=/usr/bin/mysqld_pre_systemd (code=exited, status=0/SUCCESS)
 Main PID: 13008 (mysqld)
   CGroup: /system.slice/mysqld.service
           └─13008 /usr/sbin/mysqld --daemonize --pid-file=/var/run/mysqld/mysqld.pid

5月 07 10:11:30 YYR systemd[1]: Starting MySQL Server...
5月 07 10:11:38 YYR systemd[1]: Started MySQL Server.

#将 MySQL 服务设置为开机启动
[root@YYR ~]# chkconfig mysqld on
注意:正在将请求转发到"systemctl enable mysqld.service"。
[root@YYR ~]#
```

(3) 登录 MySQL 并修改用户密码

```
# 查看 MySQL 默认密码
[root@YYR ~]# grep 'temporary password' /var/log/mysqld.log
2020-05-07T02:11:34.124784Z 1 [Note] A temporary password is generated for root@localhost: u>iKkpIyB5sn
# 登录 MySQL
[root@YYR ~]# mysql -u root -p
Enter password:
Welcome to the MySQL monitor.  Commands end with; or \g.
Your MySQL connection id is 10
Server version: 5.7.26
Copyright (c) 2000, 2020, Oracle and/or its affiliates. All rights reserved.
Oracle is a registered trademark of Oracle Corporation and/or its
affiliates. Other names may be trademarks of their respective
owners.
Type 'help;' or '\h' for help. Type '\c' to clear the current input statement.
# 临时关闭密码强度验证机制
mysql> set global validate_password_policy=0;
Query OK, 0 rows affected (0.00 sec)
# 临时关闭密码长度验证机制
mysql> set global validate_password_length=1;
Query OK, 0 rows affected (0.00 sec)
# 为 root 用户设置新的密码
mysql> alter user 'root'@'localhost' identified by 'root00';
Query OK, 0 rows affected (0.00 sec)
# 退出 MySQL
```

```
mysql >exit;
Bye
```

（4）修改 MySQL 字符集配置

```
#查看当前使用的字符集
mysql >show variables like "% char% ";
+---------------------------+---------------------------+
|Variable_name              |Value                      |
+---------------------------+---------------------------+
|character_set_client       |utf8                       |
|character_set_connection   |utf8                       |
|character_set_database     |latin1                     |
|character_set_filesystem   |binary                     |
|character_set_results      |utf8                       |
|character_set_server       |latin1                     |
|character_set_system       |utf8                       |
|character_sets_dir         |/usr/share/mysql/charsets/ |
|validate_password_special_char_count |1                |
+---------------------------+---------------------------+
9 rows in set (0.00 sec)

# 退出
mysql >exit;
Bye

# 编辑配置文件
[root@YYR ~]# vi/etc/my.cnf

# For advice on how to change settings please see
# http://dev.mysql.com/doc/refman/5.7/en/server - configuration - defaults.html

[mysqld]
#
# Remove leading # and set to the amount of RAM for the most important data
```

```
# cache in MySQL. Start at 70% of total RAM for dedicated server, else 10%.
# innodb_buffer_pool_size = 128M
#
# Remove leading # to turn on a very important data integrity option:logging
# changes to the binary log between backups.
# log_bin
#
# Remove leading # to set options mainly useful for reporting servers.
# The server defaults are faster for transactions and fast SELECTs.
# Adjust sizes as needed,experiment to find the optimal values.
# join_buffer_size = 128M
# sort_buffer_size = 2M
# read_rnd_buffer_size = 2M
datadir = /var/lib/mysql
socket = /var/lib/mysql/mysql.sock

# 增加字符集设置
character_set_server = utf8


# Disabling symbolic-links is recommended to prevent assorted security risks
symbolic-links = 0

log-error = /var/log/mysqld.log
pid-file = /var/run/mysqld/mysqld.pid

# 重启 MySQL 服务
[root@YYR ~]# service mysqld restart
Redirecting to /bin/systemctl restart mysqld.service
[root@YYR ~]# mysql -u root -p
Enter password:
Welcome to the MySQL monitor.  Commands end with;or \g.
```

```
Your MySQL connection id is 2
Server version:5.7.26 MySQL Community Server (GPL)

Copyright (c) 2000,2019,Oracle and/or its affiliates. All rights reserved.

Oracle is a registered trademark of Oracle Corporation and/or its
affiliates. Other names may be trademarks of their respective
owners.

Type 'help;' or '\h' for help. Type '\c' to clear the current input statement.

# 查看 UTF-8 字符集设置
mysql>show variables like "%char%";
+-----------------------------+-----------------------------+
|Variable_name                |Value                        |
+-----------------------------+-----------------------------+
|character_set_client         |utf8                         |
|character_set_connection     |utf8                         |
|character_set_database       |utf8                         |
|character_set_filesystem     |binary                       |
|character_set_results        |utf8                         |
|character_set_server         |utf8                         |
|character_set_system         |utf8                         |
|character_sets_dir           |/usr/share/mysql/charsets/   |
|validate_password_special_char_count |1                    |
+-----------------------------+-----------------------------+
9 rows in set (0.00 sec)

mysql>
```

任务 2　Hive 组件的新增工作

本部分的任务就是完成 Hive 组件的新增工作。

1. 安装 Hive

①从官网下载 Hive 的安装包，选择图 4 - 4 中的 “Downloads” 选项。

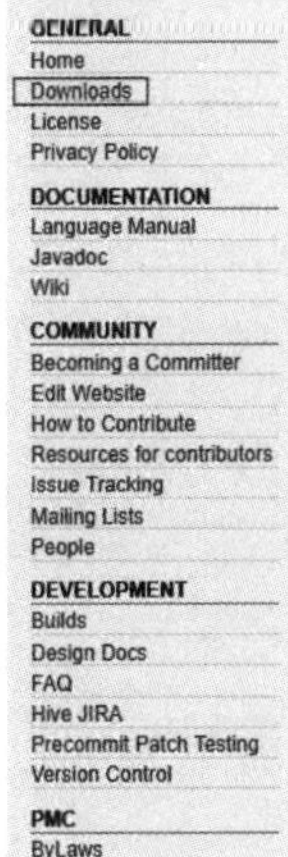

图 4 - 4　官网下载 Hive 安装包

也可以根据官方文档中提供的 URL 进行下载：http://arcHive. apache. org/dist/Hive/Hive-1. 1. 0/apache-Hive-1. 1. 0-bin. tar. gz。

②使用 SSH 工具上传 Hive 文件的压缩包到 Superintendent1 的 tmp 目录下。

③解压 Hive 文件的压缩包到/usr/local 目录中。

```
# 解压缩
[root@Superintendent1 ~]# tar -zxvf /tmp/apache-hive-1.1.0-bin.tar.gz -C /usr/local/
...(此处省略解压过程)
[root@Superintendent1 ~]# ll /usr/local/
总用量 0
drwxr-xr-x   8 root root 159 5 月   7 14:33 apache-hive-1.1.0-bin
drwxr-xr-x.  2 root root   6 11 月  5 2016 bin
drwxr-xr-x.  2 root root   6 11 月  5 2016 etc
drwxr-xr-x.  2 root root   6 11 月  5 2016 games
drwxr-xr-x  12 root root 184 4 月  28 10:02 hadoop
drwxr-xr-x.  2 root root   6 11 月  5 2016 include
drwxr-xr-x.  2 root root   6 11 月  5 2016 lib
drwxr-xr-x.  2 root root   6 11 月  5 2016 lib64
```

```
drwxr-xr-x.  2 root root   6 11月  5 2016 libexec
drwxr-xr-x.  2 root root   6 11月  5 2016 sbin
drwxr-xr-x.  5 root root  49 1月  10 10:02 share
drwxr-xr-x.  2 root root   6 11月  5 2016 src

#修改Hive目录名称
[root@Superintendent1 ~]# mv /usr/local/apache-hive-1.1.0-bin /usr/local/hive
[root@Superintendent1 ~]# ll /usr/local/
总用量 0
drwxr-xr-x.  2 root root   6 11月  5 2016 bin
drwxr-xr-x.  2 root root   6 11月  5 2016 etc
drwxr-xr-x.  2 root root   6 11月  5 2016 games
drwxr-xr-x  12 root root 184 4月  28 10:02 hadoop
drwxr-xr-x   8 root root 159 5月   7 14:33 hive
drwxr-xr-x.  2 root root   6 11月  5 2016 include
drwxr-xr-x.  2 root root   6 11月  5 2016 lib
drwxr-xr-x.  2 root root   6 11月  5 2016 lib64
drwxr-xr-x.  2 root root   6 11月  5 2016 libexec
drwxr-xr-x.  2 root root   6 11月  5 2016 sbin
drwxr-xr-x.  5 root root  49 1月  10 10:02 share
drwxr-xr-x.  2 root root   6 11月  5 2016 src

[root@Superintendent1 ~]#
```

配置环境变量

④配置环境变量。

```
#修改配置文件
[root@Superintendent1 ~]# vi /etc/profile
# /etc/profile

# System wide environment and startup programs, for login setup
# Functions and aliases go in /etc/bashrc

# It's NOT a good idea to change this file unless you know what you
# are doing. It's much better to create a custom.sh shell script in
# /etc/profile.d/ to make custom changes to your environment, as this
```

```
# will prevent the need for merging in future updates.
....(此处省略一部分内容)
# set java enviroment
export JAVA_HOME = /usr/share/java/jdk1.8.0_191
export JRE_HOME = ${JAVA_HOME}/jre
export CLASSPATH = .:${JAVA_HOME}/lib:${JRE_HOME}/lib
export PATH = $PATH:${JAVA_HOME}/bin:${JRE_HOME}/bin
# set Hadoop enviroment
export HADOOP_HOME = /usr/local/hadoop
export PATH = $PATH:${HADOOP_HOME}/sbin:${HADOOP_HOME}/bin
# 新添加内容
# set Hive enviroment
export HIVE_HOME = /usr/local/hive
export PATH = $PATH:${HIVE_HOME}/bin
# 更新环境变量
[root@Superintendent1 ~]# source/etc/profile
```

2. 配置 Hive

```
#进入 Hive 的配置文件目录 conf
[root@Superintendent1 ~]# cd/url/loca/hive/conf/
[root@Superintendent1 conf]#
#展示当前目录下的所有内容
[root@Superintendent1 conf]# ls
beeline - log4j.properties.template Hive - exec - log4j. properties.template Hive - default.xml.template Hive - env.sh.template
Hive - log4j.properties.template ivysettings.xml
#修改配置文件的名称为 Hive - site.xml,必须修改
[root@Superintendent1 conf]# mv Hive - default.xml.template Hive - site.xml
#重新查看当前目录内容,发现已经修改
[root@Superintendent1 conf]# ls
```

```
beeline-log4j.properties.template Hive-exec-log4j.properties.template Hive-site.xml
Hive-env.sh.template Hive-log4j.properties.template ivysettings.xml
#修改配置文件 Hive-site.xml,配置文件如下:
[root@Superintendent1 conf]# vi Hive-site.xml
<?xml version="1.0" encoding="UTF-8" standalone="no"?>
<?xml-stylesheet type="text/xsl" href="configuration.xsl"?><!--
Licensed to the Apache Software Foundation (ASF) under one or more
contributor license agreements. See the NOTICE file distributed with
this work for additional information regarding copyright ownership.
The ASF licenses this file to You under the Apache License, Version 2.0
(the "License"); you may not use this file except in compliance with
the License. You may obtain a copy of the License at
http://www.apache.org/licenses/LICENSE-2.0
Unless required by applicable law or agreed to in writing, software
distributed under the License is distributed on an "AS IS" BASIS,
WITHOUT WARRANTIES OR CONDITIONS OF ANY KIND, either express or implied.
See the License for the specific language governing permissions and
limitations under the License.
--><configuration>
#设置 Hive 连接 MySQL 的 URL 地址
<property>
<name>javax.jdo.option.ConnectionURL</name>
<value>jdbc:mysql://YYR:3306/Hive? createDatabaseIfNotExist=true</value>
</property>
#设置 Hive 连接 MySQL 的驱动类名称
```

```
<property>
<name>javax.jdo.option.ConnectionDriverName</name>
<value>com.mysql.cj.jdbc.Driver</value>
</property>
#设置 Hive 连接 MySQL 的用户名称
<property>
<name>javax.jdo.option.ConnectionUserName</name>
<value>root</value>
</property>
#设置 Hive 连接 MySQL 的密码
<property>
<name>javax.jdo.option.ConnectionPassword</name>
<value>root00</value>
</property>
</configuration>
#将 MySQL 的驱动包上传到/root/目录,并将此驱动包拷贝到/usrl/local/hive/lib 目录
[root@Superintendent1 ~]# cp mysql-connector-java-8.0.16.jar/usr/local/hive/lib/
```

此时，Hive 集群的 Metastore 节点已经在 Hive 上安装完成，接下来开始对 Superintendent2 客户端进行安装部署。

将 Hive 的安装包上传到 Superintendent2 节点上，对其进行解压、修改名称、配置环境变量，随后进行如下配置。

Superintendent2 上的配置如下：

```
#进入 Hive 的配置文件目录 conf
[root@Superintendent2 ~]# cd/url/loca/hive/conf/
[root@Superintendent2 conf]#
#展示当前目录下的所有内容
[root@Superintendent2 conf]# ls
beeline-log4j.properties.template Hive-exec-log4j.properties.template Hive-default.xml.template Hive-env.sh.template
Hive-log4j.properties.template ivysettings.xml
#修改配置文件的名称为 Hive-site.xml,必须修改
[root@Superintendent2 conf]# mv Hive-default.xml.template Hive-site.xml
```

```
#重新查看当前目录内容,发现已经修改
[root@Superintendent2 conf]# ls
beeline-log4j.properties.template Hive-exec-log4j.properties.template Hive-site.xml
Hive-env.sh.template Hive-log4j.properties.template ivysettings.xml
#修改配置文件 Hive-site.xml,配置文件如下:
[root@Superintendent2 conf]# vi Hive-site.xml
<?xml version="1.0" encoding="UTF-8" standalone="no"?>
<?xml-stylesheet type="text/xsl" href="configuration.xsl"?><!--
Licensed to the Apache Software Foundation (ASF) under one or more
contributor license agreements. See the NOTICE file distributed with
this work for additional information regarding copyright ownership.
The ASF licenses this file to You under the Apache License,Version 2.0
(the "License");you may not use this file except in compliance with
the License. You may obtain a copy of the License at
http://www.apache.org/licenses/LICENSE-2.0
Unless required by applicable law or agreed to in writing,software
distributed under the License is distributed on an "AS IS" BASIS,
WITHOUT WARRANTIES OR CONDITIONS OF ANY KIND,either express or implied.
See the License for the specific language governing permissions and
limitations under the License.
--><configuration>
#设置 Hive 客户端可以访问 Metastore,Hive 的 Metastore 是通过暴露 9083 端口进行连接的,因此增加如下配置
<property>
<name>Hive.metastore.uris</name>
<value>thrift://Superintendent1:9083</value>
```

```
</property>
</configuration>
```

始化元数据及排错过程

3. 初始化元数据

修改完配置文件之后，需要在 Superintendent1 中执行如下命令，进行初始化操作。也就是完成元数据存储表在 MySQL 中的表结构，包括元数据表名、元数据字段名称等信息。

```
#schematool 是初始化命令，--dbType 指定存储元数据的数据库，-initSchema 初始化表结构
[root@Superintendent1 ~]# schematool -dbType mysql -initSchema
Metastore connection URL: jdbc: Mysql:// YYR/ Hive? createDatabaseIfNotExist=true
Metastore Connection Driver:com.Mysql.cj.jdbc.Driver
Metastore connection User:root
Starting metastore schema initialization to 1.2.0
Initialization script Hive-schema-1.2.0.Mysql.sql
Initialization script completed
schemaTool completed
```

执行完成后，在 YYR 上登录 MySQL，执行下面的命令，来查看 Hive 的元数据表是否已经创建成功。

```
#在 YYR 上输入以下命令,进入 MySQL,窗口阻塞之后输入密码:123
[root@YYR local]# mysql -uroot -p
Enter password:
Welcome to the Mysql monitor. Commands end with;or \g.
Your Mysql connection id is 51
Server version:5.1.73 Source distribution
Copyright (c) 2000,2013,Oracle and/or its affiliates. All rights reserved.
Oracle is a registered trademark of Oracle Corporation and/or its
affiliates. Other names may be trademarks of their respective
owners.
Type 'help;' or '\h' for help. Type '\c' to clear the current input statement.
#登录成功
Mysql >
#展示现在存在的所有数据库
Mysql >show databases;
```

```
+--------------------+
|Database |
+--------------------+
|information_schema |
|Hive |
|Mysql |
|test |
+--------------------+
4 rows in set (0.00 sec)
#由上述结果发现,已经创建了一个叫作 Hive 的数据库,正是在配置文件中配置的数据库的名称
Mysql >
#切换数据库到 Hive
Mysql >use Hive;
Reading table information for completion of table and column names
You can turn off this feature to get a quicker startup with -A
Database changed
#展示当前 Hive 数据库下的所有表
Mysql >show tables;
+----------------------------+
|Tables_in_Hive |
+----------------------------+
|BUCKETING_COLS |
|CDS |
|COLUMNS_V2 |
|COMPACTION_QUEUE |
|COMPLETED_TXN_COMPONENTS |
|DATABASE_PARAMS |
|DBS |
|DB_PRIVS |
|DELEGATION_TOKENS |
|FUNCS |
|FUNC_RU |
|GLOBAL_PRIVS |
|HIVE_LOCKS |
|IDXS |
```

```
| INDEX_PARAMS |
| YYR_KEYS |
| NEXT_COMPACTION_QUEUE_ID |
| NEXT_LOCK_ID |
| NEXT_TXN_ID |
| NOTIFICATION LOG |
| NOTIFICATION_SEQUENCE |
| NUCLEUS_TABLES |
| PARTITIONS |
| PARTITION_EVENTS |
| PARTITION_KEYS |
| PARTITION_KEY_VALS |
| PARTITION_PARAMS |
| PART_COL_PRIVS |
| PART_COL_STATS |
| PART_PRIVS |
| ROLES |
| ROLE_MAP |
| SDS |
| SD_PARAMS |
| SEQUENCE_TABLE |
| SERDES |
| SERDE_PARAMS |
| SKEWED_COL_NAMES |
| SKEWED_COL_VALUE_LOC_MAP |
| SKEWED_STRING_LIST |
| SKEWED_STRING_LIST_VALUES |
| SKEWED_VALUES |
| SORT_COLS |
| TABLE_PARAMS |
| TAB_COL_STATS |
| TBLS |
| TBL_COL_PRIVS |
| TBL_PRIVS |
| TXNS |
| TXN_COMPONENTS |
| TYPES |
```

```
|TYPE_FIELDS |
|VERSION |
+-----------------------------+
53 rows in set (0.01 sec)
#查询到对应的表,说明 Hive 的元数据初始化工作已经完成
```

4. Hive 状态验证

上述操作结束后，Hive 的安装配置就完成了，可以通过下面的方法验证 Hive 是否安装成功。

①在 Superintendent1 上运行 hive --service metastore 命令，结果如图 4-5 所示，表明元数据启动成功。

```
[root@Slave1 local]# hive --service metastore
Starting Hive Metastore Server
```

图 4-5　在 Superintendent1 上运行 hive --service metastore 命令

IVE 状态验证及排错过程

注意：图 4-5 中的光标会阻塞住，不会出现命令提示行，这属于正常情况，所以不用担心。

②在 Superintendent2 上运行 hive 命令，结果如图 4-6 所示，表明元数据启动成功。

```
[root@Slave2 lib]# hive

Logging initialized using configuration in jar:file:/usr/local/hive-1.1.0/lib/hive-common-1.1.0.jar!/hive-log4j.propertie
s
hive>
```

图 4-6　在 Superintendent2 上运行 hive 命令

任务 3　用 Hive 来存储数据并执行查询分析

现在已经成功安装了 Hive，那么怎样运用 Hive 来存储数据并执行简单查询分析呢？首先了解 Hive 的重要概念和体系架构，然后对一个大型电商集团拥有的一批用户行为数据进行查询分析，以方便大家能够更加熟练地使用 Hive。

1. 相关概念

（1）Hive

Hive 是基于 Hadoop 的一个数据仓库工具。其本身不提供数据存储功能，使用 HDFS 存储数据。其核心工作就是把 SQL 语句翻译成 MR 程序。Hive 也不提供资源调度系统，默认由 Hadoop 中的 YARN 集群来调度，可以将结构化的数据映射为一张数据库表，并提供 HQL 查询功能。

（2）企业使用 Hive 的利弊

在企业中直接使用 Hadoop，将会导致人员学习成本高、项目周期要求短、MapReduce 实现复杂查询逻辑开发难度大等问题。

但是 Hive 在操作上的优势使企业对其难以抗拒：①操作接口采用类 SQL 语法，提供快速开发的能力；②避免了写 MapReduce，减少开发人员的学习成本；③可以很方便地扩展功能。

（3）特点

①简单：提供了类 SQL 查询语言 HQL。

②可扩展：为超大数据集设计了计算/扩展能力，一般情况下不需要重启服务。Hive 可以自由地扩展集群的规模。

③便于管理：提供统一的元数据管理。

④延展性：Hive 支持用户自定义函数，用户可以根据自己的需求来实现自己的函数。

⑤容错性：良好的容错性。节点出现问题时，SQL 仍可完成执行。

（4）Hive 的 Metastore

Hive 的元数据（Metastore）包含用 Hive 创建的 database、table 等元信息。元数据存储在关系型数据库中，如 Derby、MySQL 等，尽量不要使用 Derby。

Hive 的 Metastore 是客户端用来连接 Metastore 服务的，Metastore 再去连接 MySQL 数据库来存取元数据。有了 Metastore 服务后，就可以有多个客户端同时连接，并且这些客户端不需要知道 MySQL 数据库的用户名和密码，只需要连接 Metastore 服务即可。

2. Hive 体系结构

（1）Hive 的架构

Hive 作为企业级的数据仓库，其架构也是非常重要的，如图 4－7 所示。

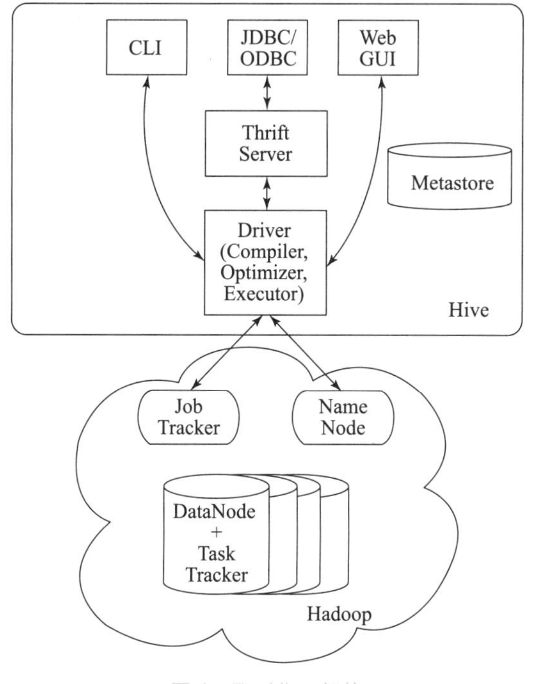

图 4－7　Hive 架构

HIVE 体系架构

从 Hive 的架构图可以看出，其数据的存储是依托于 HDFS 的。此外，Hive 还包含三个最核心的组件：用户接口、Metastore、Driver。

Hive 的体系结构可以分为以下几部分：

①用户接口，主要有三个：CLI、Client 和 WUI。其中最常用的是 CLI。CLI 启动的时候，会同时启动一个 Hive 副本。Client 是 Hive 的客户端，用户将其连接至 Hive Server。在启动 Client 的时候，需要给出 Hive Server 所在节点，并且在该节点启动 Hive Server。WUI 是通过浏览器来对 Hive 进行访问的。

②元数据，存储在数据库中，如 MySQL、Derby。Hive 中的元数据包括表的名字、表的列和分区及其属性、表的数据所在目录等。

③解释器、编译器、优化器，完成 HQL 查询语句从词法分析、语法分析、编译、优化及查询计划的生成。生成的查询计划存储在 HDFS 中，并在 MapReduce 中调用执行。

④数据，存储在 HDFS 中，大部分的查询、计算由 MapReduce 完成（包含 * 的查询，比如 select * from tbl 不会生成 MapRedcue 任务）。

（2）Hive 的工作原理

图 4－8 所示为 Hive 与 Hadoop 之间的工作流程。

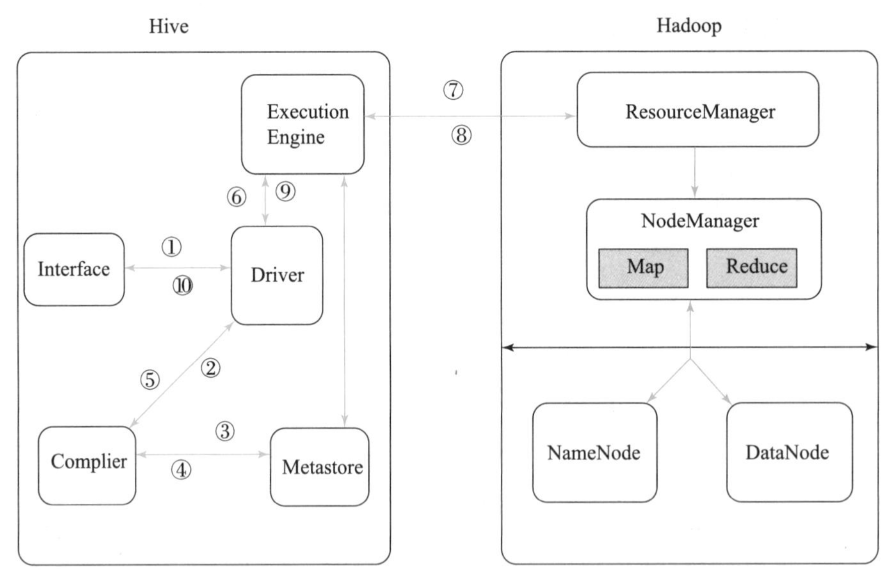

图 4－8 Hive 与 Hadoop 之间的工作流程

①Interface：指的是 Hive 的用户接口，如命令行或者 Web UI 发送查询语句来执行操作。

②将 SQL 语句提交到 Hive 的 Driver，在驱动程序的帮助下查询编译器，分析查询语法和查询计划或者查询的要求。

③编译器发送元数据请求到 Metastore，查看需要的表名、字段名等是否存在。

④Metastore 响应编译器发送到元数据的请求，如果满足，则接着向下执行，否则，将错误信息返回给用户接口。

⑤编译器重新检查要求，并重新发送计划给驱动程序，此时查询解析和编译的工作完成。

⑥驱动程序发送执行计划到执行引擎。

⑦执行作业的过程是一个 MapReduce 工作，执行引擎发送作业给 ResourceManager，同时名称节点分配作业到 NodeManager，在此，查询执行 MapReduce 工作。此外，执行引擎可通过 Metastore 执行元数据操作。

⑧执行引擎接收来自数据节点的结果。

⑨执行引擎发送结果给驱动程序。

(3) 与传统关系型数据库的区别

通过表 4-2 来比较 Hive 与传统关系型数据库中表的区别。

表 4-2　Hive 与传统关系型数据库中表的区别

对比项	Hive	RDRMS
查询语言	HQL	SQL
数据存储	HDFS	Raw Device 或 Local FS
执行器	MapReduce	Executor
数据插入	支持批量导入/单条插入	支持批量导入/单条插入
数据操作	覆盖追加	行级更新删除
处理数据规模	大	小
执行延迟	高	低
分区	支持	支持
索引	0.8 版本之后加入简单索引	支持复杂的索引
扩展性	高（好）	有限（差）
数据加载模式	读时模式（快）	写时模式（慢）
应用场景	海量数据查询	实时查询

3. 应用案例

某个大型电商集团拥有一批用户行为数据，那么如何用 Hive 来存储数据并执行一些简单的查询分析呢？下面来了解一下这个项目。

(1) 需求背景

随着科学技术的发展，人们在网上购物的需求越来越大，某个电商集团积累了一批用户行为数据，需要在这些数据的基础上进行某些分析查询。

(2) 数据字典（表 4-3）

表4-3 数据字典

ID	说明
User_id	用户 ID
Item_id	商品 ID
behavior_type	用户行为类型（浏览、收藏、加购物车、购买分别对应1、2、3、4）
item_category	商品类别
visit_date	时间
province	省份

（3）样例数据（图4-9）

```
299943  104078656  218677865  1  2993   2014-12-04  云南
299944  104078656  265098764  1  3628   2014-11-27  新疆
299945  104078656  365292426  1  2901   2014-12-09  内蒙古
299946  104078656  155767031  1  11552  2014-12-03  山东
299947  104078656  49163285   1  3892   2014-11-19  广西
299948  104078656  166487641  1  2993   2014-12-08  新疆
299949  104078656  386172168  1  11552  2014-11-23  浙江
299950  104078656  371246765  1  7995   2014-12-03  云南
299951  104078656  27582949   1  11552  2014-12-11  香港
299952  104078656  389420394  1  8537   2014-12-07  辽宁
299953  104078656  88472678   1  3628   2014-12-01  江西
299954  104078656  227864974  1  3628   2014-12-09  安徽
299955  104078656  267358911  1  12488  2014-11-26  宁夏
```

图4-9 样例数据

（4）提前准备好数据文件，将数据文件上传到HDFS集群，执行以下命令

```
1.#在 HDFS 中创建目录
2.[root@Superintendent1 ~]# hdfs dfs -mkdir -p/bigdatacase/data-
  set
3.#上传文件到 HDFS 中刚刚创建的文件夹下
4.[root@Superintendent1 ~]# hdfs dfs -put/tmp/user_table.txt/
  bigdatacase/dataset
5.#查看刚刚上传文件的前10行
6.[root@Superintendent1 ~]# hdfs dfs -cat/bigdatacase/dataset/
  user_table.txt |head -10
7.1 10001082 285259775 1 4076 2014-12-08 辽宁
8.2 10001082 4368907 1 5503 2014-12-12 黑龙江
9.3 10001082 4368907 1 5503 2014-12-12 山东
10.4 10001082 53616768 1 9762 2014-12-02 山西
11.5 10001082 151466952 1 5232 2014-12-12 海南
12.6 10001082 53616768 4 9762 2014-12-02 上海市
13.7 10001082 290088061 1 5503 2014-12-12 青海
```

```
8 10001082 298397524 1 10894 2014-12-12 澳门
9 10001082 32104252 1 6513 2014-12-12 广东
10 10001082 323339743 1 10894 2014-12-12 湖北
```

（5）在 Superintendent2 的 Hive 窗口中创建 Hive 运行需要的表

```
Hivc > CREATE EXTERNAL TABLE bigdata_user(id INT,uid STRING,item_id STRING,behavior_type INT,item_category STRING,visit_date DATE,province STRING) ROW FORMAT DELIMITED FIELDS TERMINATED BY '\t' STORED AS TEXTFILE LOCATION '/bigdatacase/dataset';
OK;Time taken:0.205 seconds
Hive >
```

（6）SQL 查询

下面将根据上面创建完成的表进行一系列的查询。

①查看前 10 位用户对商品的行为。

```
Hive >select behavior_type from bigdata_user limit 10;
OK
1
1
1
1
1
4
1
1
1
1
Time taken:0.105 seconds,Fetched:10 row(s)
```

②查询前 20 位用户购买商品的时间和商品的种类。

```
Hive >select visit_date,item_category from bigdata_user limit 20;
OK
2014-12-08 4076
2014-12-12 5503
2014-12-12 5503
2014-12-02 9762
2014-12-12 5232
```

```
2014-12-02 9762
2014-12-12 5503
2014-12-12 10894
2014-12-12 6513
2014-12-12 10894
2014-12-12 2825
2014-11-28 2825
2014-12-15 3200
2014-12-03 10576
2014-11-20 10576
2014-12-13 10576
2014-12-08 10576
2014-12-14 7079
2014-12-02 6669
2014-12-12 5232
Time taken:0.095 seconds,Fetched:20 row(s)
```

③计算表内数据行数。

```
Hive>select count(*) from bigdata_user;
Query ID = root_20200329045812_d54786a1-37f9-499e-ba90-f39c8930d7c1
Total jobs=1
Launching Job 1 out of 1
Number of reduce tasks determined at compile time:1
In order to change the average load for a reducer (in bytes):
set Hive.exec.reducers.bytes.per.reducer=<number>
In order to limit the maximum number of reducers:
set Hive.exec.reducers.max=<number>
In order to set a constant number of reducers:
set MapReduce.job.reduces=<number>
Starting Job=job_1553806653852_0001,Tracking URL=http://YYR:8088/proxy/application_1553806653852_0001/
Kill Command=/root/Hadoop-2.6.5/bin/Hadoop job -kill job_1553806653852_0001
Hadoop job information for Stage-1:number of mappers:1;number of reducers:1
2020-03-29 04:58:28,654 Stage-1 map=0%,reduce=0%
```

```
2020-03-29 04:58:40,687 Stage-1 map=100%,reduce=0%,Cumulative CPU 1.66 sec
2020-03-29 04:58:49,075 Stage-1 map=100%,reduce=100%,Cumulative CPU 3.41 sec
MapReduce Total cumulative CPU time:3 seconds 410 msec
Ended Job=job_1553806653852_0001
MapReduce Jobs Launched:
Stage-Stage-1:Map:1 Reduce:1 Cumulative CPU:3.41 sec HDFS Read:15598233 HDFS Write:7 SUCCESS
Total MapReduce CPU Time Spent:3 seconds 410 msec
OK
300000
Time taken:37.468 seconds,Fetched:1 row(s)
```

④查询用户不重复的数据。

```
Hive>select count(distinct uid) from bigdata_user;
Query ID=root_20200329045948_e8317f00-be76-45d7-8633-9f45d3443565
Total jobs=1
Launching Job 1 out of 1
Number of reduce tasks determined at compile time:1
In order to change the average load for a reducer (in bytes):
set Hive.exec.reducers.bytes.per.reducer=<number>
In order to limit the maximum number of reducers:
set Hive.exec.reducers.max=<number>
In order to set a constant number of reducers:
set MapReduce.job.reduces=<number>
Starting Job=job_1553806653852_0002,Tracking URL=http://YYR:8088/proxy/application_1553806653852_0002/
Kill Command=/root/Hadoop-2.6.5/bin/Hadoop job -kill job_1553806653852_0002
Hadoop job information for Stage-1:number of mappers:1;number of reducers:1
2020-03-29 05:00:01,495 Stage-1 map=0%,reduce=0%
2020-03-29 05:00:10,047 Stage-1 map=100%,reduce=0%,Cumulative CPU 1.88 sec
```

```
2020 -03 -29 05:00:18,384 Stage -1 map =100% ,reduce =100% ,Cumulative CPU 3.74 sec
MapReduce Total cumulative CPU time:3 seconds 740 msec
Ended Job = job_1553806653852_0002
MapReduce Jobs Launched:
Stage - Stage - 1: Map: 1 Reduce: 1 Cumulative CPU: 3.74 sec HDFS Read:15598238 HDFS Write:4 SUCCESS
Total MapReduce CPU Time Spent:3 seconds 740 msec
OK
270
Time taken:31.15 seconds,Fetched:1 row(s)
```

⑤查询不重复的数据有多少条。

```
Hive >select count( * ) from (select uid,item_id,behavior_type,item_category,visit_date,province from bigdata_user group by
uid,item_id, behavior_type, item_category, visit_date,province having count( * ) =1)a;
Query ID = root _20200329050051 _77799ec5 - 9c8d - 4b71 - 924d - b04b3f6486f9
Total jobs =2
Launching Job 1 out of 2
Number of reduce tasks not specified. Estimated from input data size:1
In order to change the average load for a reducer (in bytes):
set Hive. exec. reducers. bytes. per. reducer = <number >
In order to limit the maximum number of reducers:
set Hive. exec. reducers. max = <number >
In order to set a constant number of reducers:
set MapReduce. job. reduces = <number >
Starting Job = job_1553806653852_0003,Tracking URL = http://YYR:8088/proxy/application_1553806653852_0003/
Kill Command = /root/Hadoop - 2.6.5/bin/Hadoop job - kill job_1553806653852_0003
Hadoop job information for Stage -1:number of mappers:1;number of reducers:1
2020 -03 -29 05:01:00,452 Stage -1 map =0% ,reduce =0%
```

```
2020 -03 -29 05:01:14,019 Stage -1 map =100% ,reduce =0% ,Cumulative CPU 6.02 sec
2020 -03 -29 05:01:27,606 Stage -1 map =100% ,reduce =100% ,Cumulative CPU 6.02 sec
MapReduce Total cumulative CPU time:6 seconds 20 msec
Ended Job = job_1553806653852_0003
Launching Job 2 out of 2
Number of reduce tasks determined at compile time:1
In order to change the average load for a reducer (in bytes):
set Hive.exec.reducers.bytes.per.reducer = <number>
In order to limit the maximum number of reducers:
set Hive.exec.reducers.max = <number>
In order to set a constant number of reducers:
set MapReduce.job.reduces = <number>
Starting Job = job_1553806653852_0004,Tracking URL = http://YYR:8088/proxy/application_1553806653852_0004/
Kill Command = /root/Hadoop -2.6.5/bin/Hadoop job - kill job_1553806653852_0004
Hadoop job information for Stage -2:number of mappers:1;number of reducers:1
2020 -03 -29 05:01:38,015 Stage -2 map =0% ,reduce =0%
2020 -03 -29 05:01:46,346 Stage -2 map =100% ,reduce =0% ,Cumulative CPU 1.08 sec
2020 -03 -29 05:01:54,742 Stage -2 map =100% ,reduce =100% ,Cumulative CPU 2.86 sec
MapReduce Total cumulative CPU time:2 seconds 860 msec
Ended Job = job_1553806653852_0004
MapReduce Jobs Launched:
Stage - Stage -1: Map:1 Reduce:1 Cumulative CPU:11.17 sec HDFS Read:15600704 HDFS Write:117 SUCCESS
Stage - Stage -2: Map:1 Reduce:1 Cumulative CPU:2.86 sec HDFS Read:4455 HDFS Write:7 SUCCESS
Total MapReduce CPU Time Spent:14 seconds 30 msec
OK
284181
Time taken:64.171 seconds,Fetched:1 row(s)
```

运行完成后，Hive组件运行正常。

学习笔记

项目五
Flume 的应用

项目描述

假设需要通过爬虫系统从各大招聘网站上爬取对应职位信息，将爬取的信息保存到本地文件中，并进行一系列预处理。预处理后，需要将数据上传到 HDFS 上进行统一存储和分析，可以使用 Flume 工具实现。

项目分析

通过以下几个任务来完成对知识点 Flume 的学习：

任务 1：Flume 组件的安装。

任务 2：Flume 的运行机制。

任务 3：Flume 的应用案例。

Cloudera 提供的 Flume 是一个高可用、高可靠、分布式的海量日志采集、聚合和传输的系统。Flume 支持在日志系统中定制各类数据发送方，用来收集数据；同时，提供对数据进行简单处理，并写到各种数据接收方（可定制）。

Flume OG（Original Generation）是 Flume 初始的发行版本，属于 Cloudera。随着 Flume 功能的扩展，Flume OG 代码工程臃肿、核心组件设计不合理、核心配置不标准等缺点逐渐暴露，尤其是在 Flume OG 的最后一个发行版本 0.94.0 中，日志传输不稳定现象非常严重。为解决这些问题，2011 年 10 月 22 日，Cloudera 完成了 Flume-728，对 Flume 进行了里程碑式的改动：重构核心组件、核心配置及代码架构，重构后的版本统称为 Flume NG（Next Generation）。改动的另一个主要原因是 Flume 被纳入 Apache 旗下，Cloudera Flume 更名为 Apache Flume。

任务 1　Flume 组件的安装

下载 FLUME 安装包

1. 安装 Flume

①从官方网站下载 Flume 安装包，单击图 5-1 中“Download”选项，选择相应的 Flume

版本进行安装。

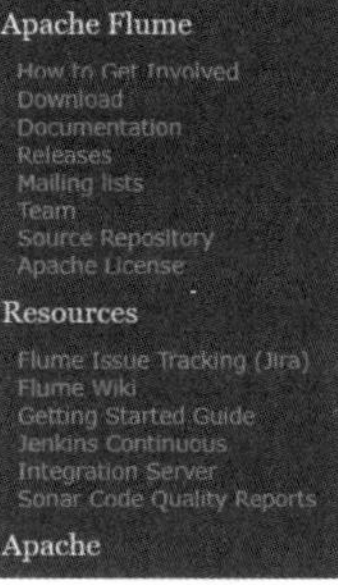

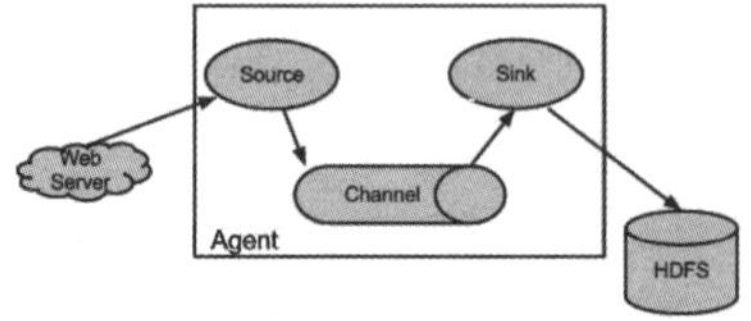

图 5－1　官网站下载 Flume 安装包

②下载完成后，上传 Flume 的压缩文件到 Superintendent2 中。

③解压 Flume 的压缩文件到/usr/local/目录下。

```
[root@Superintendent2 ~]# tar -zxvf apache-flume-1.6.0-bin.tar.gz -C /usr/local/
#查看解压后的目录文件
[root@Superintendent2 ~]# ls /usr/local/
apache-flume-1.6.0-bin bin etc games hadoop-2.6.0 hive-1.1.0 include lib lib64 libexec sbin share src
```

④修改 Flume 文件夹的名称为 flume－1.6.0。

```
[root@Superintendent2 local]# mv /usr/local/apache-flume-1.6.0-bin/ /usr/local/flume
#查看修改后的/usr/local 文件夹下的内容
[root@Superintendent2 local]# ls /usr/local
bin etc flume games hadoop hive include lib lib64 libexec sbin share src
```

⑤配置 Flume 的环境变量。

```
#打开配置文件
[root@Superintendent2 local]# vi /etc/profile
#跳转到文件的最后一行,修改为如下配置
export FLUME_HOME=/usr/local/flume
export PATH=$PATH:${FLUME_HOME}/bin
```

配置 FLUME 环境变量

⑥执行命令，让刚刚修改的环境变量生效。

```
[root@Superintendent2 local]# source/etc/profile
```

2. 验证 Flume 安装是否成功

在 Superintendent2 上运行 flume - ng help 命令，出现 Flume 命令的帮助手册。

```
#查看 Flume 的帮助手册
[root@Superintendent2 ~]# flume -ng help
Usage:/usr/local/flume/bin/flume -ng <command>[options]...
commands:
help display this help text
agent run a Flume agent
avro -client run an avro Flume client
version show Flume version info
global options:
 --conf, -c <conf>use configs in <conf>directory
 --classpath, -C <cp>append to the classpath
 --dryrun, -d do not actually start Flume,just print the command
 --plugins -path <dirs>colon -separated list of plugins.d directories. See the
plugins.d section in the user guide for more details.
Default: $FLUME_HOME/plugins.d
 -Dproperty =value sets a Java system property value
 -Xproperty =value sets a Java -X option
agent options:
 --name, -n <name>the name of this agent (required)
 --conf -file, -f <file>specify a config file (required if -z missing)
 --zkConnString, -z <str>specify the ZooKeeper connection to use (required if -f missing)
 --zkBasePath, -p <path>specify the base path in ZooKeeper for agent configs
 --no -reload -conf do not reload config file if changed
 --help, -h display help text
avro -client options:
 --rpcProps, -P <file>RPC client properties file with server connection params
```

验证 FLUME 安装是否成功

```
--host, -H <host> hostname to which events will be sent
--port, -p <port> port of the avro source
--dirname <dir> directory to stream to avro source
-- filename, - F < file > text file to stream to avro source (default:std input)
--headerFile, -R <file> File containing event headers as key/value pairs on each new line
--help, -h display help text
Either --rpcProps or both --host and --port must be specified.
Note that if <conf> directory is specified,then it is always included first
in the classpath.
```

运行 flume-ng version 命令，查看 Flume 的版本信息。如果显示 Flume 的版本信息，则表示安装成功，如图 5-2 所示；反之，则安装失败。

```
[root@Slave2 ~]# flume-ng version
Flume 1.6.0
Source code repository: https://git-wip-us.apache.org/repos/asf/flume.git
Revision: 2561a23240a71ba20bf288c7c2cda88f443c2080
Compiled by hshreedharan on Mon May 11 11:15:44 PDT 2015
From source with checksum b29e416802ce9ece3269d34233baf43f
```

图 5-2　Flume 安装成功

任务 2　Flume 的运行机制

任务 1 中成功安装了 Flume，现在需要对 Flume 的一些重要概念、体系架构进行更加深入的了解，以方便在项目中合理地选择需要的组件，达到更快捷、高效的数据传输的目的。

1. Flume 的相关概念

Agent 完成的功能是从一个 WebServer 读取到对应的日志数据，将数据写入 HDFS 中。在 Flume 的官网 http://flume.apache.org/可以看到 Flume 的每一个 Agent 的组件架构，如图 5-3 所示。其中，Source、Channel、Sink 是核心组件。

Client：

Client 主要用来生产数据，其运行在一个独立的线程上，图中的 WebServer 就是一个Client。

Event：

一个数据单元，由消息头和消息体组成（Event 可以是日志记录、avro 对象等。）

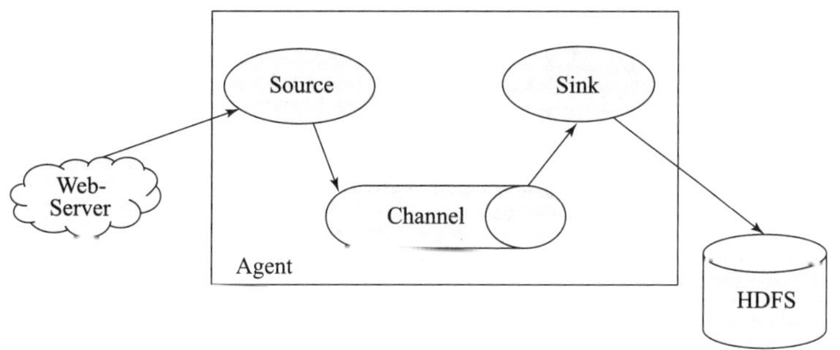

图 5-3 Agent 的组件架构图

Source：

数据收集组件（Source 从 Client 收集数据，传递给 Channel）。

Channel：

临时存储数据，用来中转 Event 的临时信息，保存由 Source 组件传递过来的 Event。

Sink：

从 Channel 中读取并移除 Event，将 Event 传递到接收数据的目的源。

通过学习，发现其实 Flume 的架构并不复杂，在使用过程中最主要的是要了解 Flume 可以从哪些数据源接收数据，可以使用哪些缓存来暂存数据，又可以将数据发送到什么样的目的源。

2. Flume 的 source

在 Flume 的官网中看到表 5-1 所示的几种 Source。

表 5-1 常见 Source

Avro Source	支持 Avro 协议，实际是 Avro RPC
Thrift Source	支持 Thrift 协议
Exec Source	基于 UNIX 的 command 在标准输出上生产数据
JMS Source	从 JMS 中读取数据
Spooling Directory	监控指定目录内的数据变更
Netcat Source	监控某个端口，将流经端口的每一个文本行数据作为输入
Sequence Generator	序列生成器数据源，生产序列数据
Syslog Source	读取 Syslog 数据，支持 TCP 和 UDP 数据

续表

HTTP Source	基于 HTTP POST 和 GET 的数据源，支持 JSON 和 BLOB
Kafka Source	从 Kafka 中读取数据
Legacy Source	兼容老的 Flume OG 中的 Source

这里介绍主要的几个 Source 源：

（1）Avro Source

监听 Avro 端口，从 Avro Client Streams 接收数据，参数列表如图 5-4 所示。

Property Name	Default	Description
channels	–	
type	–	The component type name, needs to be thrift
bind	–	hostname or IP address to listen on
port	–	Port # to bind to
threads	–	Maximum number of worker threads to spawn
selector.type		
selector.*		
interceptors	–	Space separated list of interceptors
interceptors.*		
ssl	false	Set this to true to enable SSL encryption. You must also specify a "keystore" and a "keystore-password".
keystore	–	This is the path to a Java keystore file. Required for SSL.
keystore-password	–	The password for the Java keystore. Required for SSL.
keystore-type	JKS	The type of the Java keystore. This can be "JKS" or "PKCS12".
exclude-protocols	SSLv3	Space-separated list of SSL/TLS protocols to exclude. SSLv3 will always be excluded in addition to the protocols specified.
kerberos	false	Set to true to enable kerberos authentication. In kerberos mode, agent-principal and agent-keytab are required for successful authentication. The Thrift source in secure mode, will accept connections only from Thrift clients that have kerberos enabled and are successfully authenticated to the kerberos KDC.
agent-principal	–	The kerberos principal used by the Thrift Source to authenticate to the kerberos KDC.
agent-keytab	—	The keytab location used by the Thrift Source in combination with the agent-principal to authenticate to the kerberos KDC.

图 5-4　Avro Source 参数列表

具体使用方式如下：

```
1.#指定 Source 的名称
2.a1.sources = r1
3.#指定 Channel 的名称
4.a1.channels = c1
5.#指定 Source 的类型
6.a1.sources.r1.type = avro
7.#将 Source 跟 Channel 连接起来
8.a1.sources.r1.channels = c1
9.#配置 Avro Source 源绑定的主机名称
10.a1.sources.r1.bind = 0.0.0.0
```

```
#指定 Avro Source 源绑定主机的端口号
a1.sources.r1.port =4141
```

（2）Thrift Source

监听 Thrift 端口，从 Thrift Client Streams 接收数据。参数列表如图 5－5 所示。

Property Name	Default	Description
channels	–	
type	–	The component type name, needs to be avro
bind	–	hostname or IP address to listen on
port	–	Port # to bind to
threads	–	Maximum number of worker threads to spawn
selector.type		
selector.*		
interceptors	–	Space-separated list of interceptors
interceptors.*		
compression-type	none	This can be "none" or "deflate". The compression-type must match the compression-type of matching AvroSource
ssl	false	Set this to true to enable SSL encryption. You must also specify a "keystore" and a "keystore-password".
keystore	–	This is the path to a Java keystore file. Required for SSL.
keystore-password	–	The password for the Java keystore. Required for SSL.
keystore-type	JKS	The type of the Java keystore. This can be "JKS" or "PKCS12".
exclude-protocols	SSLv3	Space-separated list of SSL/TLS protocols to exclude. SSLv3 will always be excluded in addition to the protocols specified.
ipFilter	false	Set this to true to enable ipFiltering for netty
ipFilterRules	–	Define N netty ipFilter pattern rules with this config.

图 5－5　Thrift Source 参数列表

具体使用方式如下：

```
#指定 Source 的名称
a1.sources = r1
#指定 Channel 的名称
a1.channels = c1
#指定 Source 的类型
a1.sources.r1.type = thrift
#将 Source 跟 Channel 连接起来
a1.sources.r1.channels = c1
#配置 Avro Source 源绑定的主机名称
a1.sources.r1.bind =0.0.0.0
#指定 Avro Source 源绑定主机的端口号
a1.sources.r1.port =4141
```

（3）Exec Source

Exec Source 的配置就是设定一个 Linux 命令，然后通过这个命令不断输出数据。如果进程退出，Exec Source 也一起退出，不会产生进一步的数据，参数列表如图 5－6 所示。

Property Name	Default	Description
channels	–	
type	–	The component type name, needs to be exec
command	–	The command to execute
shell	–	A shell invocation used to run the command. e.g. /bin/sh -c. Required only for commands relying on shell features like wildcards, back ticks, pipes etc.
restartThrottle	10000	Amount of time (in millis) to wait before attempting a restart
restart	false	Whether the executed cmd should be restarted if it dies
logStdErr	false	Whether the command's stderr should be logged
batchSize	20	The max number of lines to read and send to the channel at a time
batchTimeout	3000	Amount of time (in milliseconds) to wait, if the buffer size was not reached, before data is pushed downstream
selector.type	replicating	replicating or multiplexing
selector.*		Depends on the selector.type value
interceptors	–	Space-separated list of interceptors
interceptors.*		

图 5-6　Exec Source 参数列表

具体的使用方法如下：

```
#指定 Source 的名称
a1.sources = r1
#指定 Channel 的名称
a1.channels = c1
#指定 Source 的类型
a1.sources.r1.type = exec
#将 Source 跟 Channel 连接起来
a1.sources.r1.channels = c1
#配置 Exec Source 使用的 UNIX 命令
a1.sources.r1.command = tail -F /var/log/secure
```

（4）Spooling Directory Source

Spooling Directory Source 监测配置的目录下新增的文件，并将文件中的数据读取出来。但是，Spooling Directory Source 有两个需要注意的地方：第一个是拷贝到 spool 目录下的文件不能对其进行再编辑，第二个是 spool 目录下不能包含对应的子目录。Spooling Directory Source 主要用来对日志进行准实时监控。参数列表如图 5-7 所示。

具体使用方式如下：

```
#指定 Source 的名称
a1.sources = r1
#指定 Channel 的名称
a1.channels = c1
#指定 Source 的类型
a1.sources.r1.type = spooldir
#将 Source 跟 Channel 连接起来
a1.sources.r1.channels = c1
```

Property Name	Default	Description
channels	–	
type	–	The component type name, needs to be spooldir.
spoolDir	–	The directory from which to read files from.
fileSuffix	.COMPLETED	Suffix to append to completely ingested files
deletePolicy	never	When to delete completed files: never or immediate
fileHeader	false	Whether to add a header storing the absolute path filename.
fileHeaderKey	file	Header key to use when appending absolute path filename to event header.
basenameHeader	false	Whether to add a header storing the basename of the file.
basenameHeaderKey	basename	Header Key to use when appending basename of file to event header.
ignorePattern	^$	Regular expression specifying which files to ignore (skip)
trackerDir	.flumespool	Directory to store metadata related to processing of files. If this path is not an absolute path, then it is interpreted as relative to the spoolDir.
consumeOrder	oldest	In which order files in the spooling directory will be consumed oldest, youngest and random. In case of oldest and youngest, the last modified time of the files will be used to compare the files. In case of a tie, the file with smallest laxicographical order will be consumed first. In case of random any file will be picked randomly. When using oldest and youngest the whole directory will be scanned to pick the oldest/youngest file, which might be slow if there are a large number of files, while using random may cause old files to be consumed very late if new files keep coming in the spooling directory.
maxBackoff	4000	The maximum time (in millis) to wait between consecutive attempts to write to the channel(s) if the channel is full. The source will start at a low backoff and increase it exponentially each time the channel throws a ChannelException, upto the value specified by this parameter.
batchSize	100	Granularity at which to batch transfer to the channel
inputCharset	UTF-8	Character set used by deserializers that treat the input file as text.
decodeErrorPolicy	FAIL	What to do when we see a non-decodable character in the input file. FAIL: Throw an exception and fail to parse the file. REPLACE: Replace the unparseable character with the "replacement character" char, typically Unicode U+FFFD. IGNORE: Drop the unparseable character sequence.
deserializer	LINE	Specify the deserializer used to parse the file into events. Defaults to parsing each line as an event. The class specified must implement EventDeserializer.Builder.
deserializer.*		Varies per event deserializer.
bufferMaxLines	–	(Obselete) This option is now ignored.
bufferMaxLineLength	5000	(Deprecated) Maximum length of a line in the commit buffer. Use deserializer.maxLineLength instead.
selector.type	replicating	replicating or multiplexing
selector.*		Depends on the selector.type value
interceptors	–	Space-separated list of interceptors
interceptors.*		

图5-7　Spooling Directory Source 参数列表

```
#配置 Spooling Directory Source 要监控的目录
a1.sources.r1.spoolDir = /var/log/apache/flumeSpool
#将文件头信息(来自哪个文件)添加到输出数据中
a1.sources.r1.fileHeader = true
```

(5) Netcat Source

Netcat Source 在某一端口上进行侦听，它将每一行文字变成一个事件源，也就是数据通过换行符进行分隔。它的工作就像命令 nc -k -l[host][port]。或者理解为，它打开一个指定端口，侦听数据，将每一行文字变成 Flume 事件，并通过连接通道发送。参数列表如图5-8所示。

具体使用方式如下：

```
#指定 Source 的名称
a1.sources = r1
#指定 Channel 的名称
a1.channels = c1
#指定 Source 的类型
```

Property Name	Default	Description
channels	–	
type	–	The component type name, needs to be netcat
bind	–	Host name or IP address to bind to
port	–	Port # to bind to
max-line-length	512	Max line length per event body (in bytes)
ack-every-event	true	Respond with an "OK" for every event received
selector.type	replicating	replicating or multiplexing
selector.*		Depends on the selector.type value
interceptors	–	Space-separated list of interceptors
interceptors.*		

图 5－8　Netcat Source 参数列表

```
a1.sources.r1.type = netcat
#将 Source 跟 Channel 连接起来
a1.sources.r1.channels = c1
#配置 Netcat Source 绑定主机的名称
a1.sources.r1.bind = 0.0.0.0
#配置 Netcat Source 绑定主机的端口号
a1.sources.r1.port = 6666
```

（6） HTTP Source

HTTP POST 和 GET 是提交数据的两种方式，官网说 GET 应只用于实验。Flume 事件使用一个可插拔的 "handler" 程序来实现数据之间的转换，它必须实现 HTTPSourceHandler 接口。HTTP Source 处理程序需要 HttpServletRequest 和 Flume 来支持，参数列表如图 5－9 所示。

Property Name	Default	Description
type		The component type name, needs to be http
port	–	The port the source should bind to.
bind	0.0.0.0	The hostname or IP address to listen on
handler	org.apache.flume.source.http.JSONHandler	The FQCN of the handler class.
handler.*	–	Config parameters for the handler
selector.type	replicating	replicating or multiplexing
selector.*		Depends on the selector.type value
interceptors	–	Space-separated list of interceptors
interceptors.*		
enableSSL	false	Set the property true, to enable SSL. *HTTP Source does not support SSLv3.*
excludeProtocols	SSLv3	Space-separated list of SSL/TLS protocols to exclude. SSLv3 is always excluded.
keystore		Location of the keystore includng keystore file name
keystorePassword	Keystore password	

图 5－9　HTTP Source 参数列表

具体使用方式如下：

```
#指定 Source 的名称
a1.sources = r1
```

```
#指定 Channel 的名称
a1.channels = c1
#指定 Source 的类型
a1.sources.r1.type = http
#将 Source 跟 Channel 连接起来
a1.sources.r1.channels = c1
#配置 HTTP Source 的端口号
a1.sources.r1.port = 5140
#配置 HTTP Source 的处理类
a1.sources.r1.handler = org.example.rest.RestHandler
#配置 HTTP Source 的处理类的参数
a1.sources.r1.handler.nickname = random props
```

（7） Kafka Source

Kafka Source 是一个从 Kafka 的 topic 中读取消息的 Apache Kafka 的消费者。如果运行着许多的 Kafka Source，可以使用相同的消费者组来配置它们，以保证它们读到 topic 中唯一分区集中的数据。参数列表如图 5－10 所示。

Property Name	Default	Description
channels	–	
type	–	The component type name, needs to be org.apache.flume.source.kafka.KafkaSource
zookeeperConnect	–	URI of ZooKeeper used by Kafka cluster
groupId	flume	Unique identified of consumer group. Setting the same id in multiple sources or agents indicates that they are part of the same consumer group
topic	–	Kafka topic we'll read messages from. At the time, this is a single topic only.
batchSize	1000	Maximum number of messages written to Channel in one batch
batchDurationMillis	1000	Maximum time (in ms) before a batch will be written to Channel The batch will be written whenever the first of size and time will be reached.
Other Kafka Consumer Properties	–	These properties are used to configure the Kafka Consumer. Any producer property supported by Kafka can be used. The only requirement is to prepend the property name with the prefix kafka.. For example: kafka.consumer.timeout.ms Check *Kafka documentation* *<https://kafka.apache.org/08/configuration.html#consumerconfigs>* for details

图 5－10　Kafka Source 参数列表

具体使用方式如下：

```
#指定 Source 的名称
a1.sources = r1
#指定 Channel 的名称
a1.channels = c1
#指定 Source 的类型
a1.sources.r1.type = org.apache.flume.source.kafka.KafkaSource
#将 Source 和 Channel 连接起来
a1.sources.r1.channels = c1
#指定 Kafka 集群 ZooKeeper 的地址
```

```
a1.sources.r1.zookeeperConnect = localhost:2181
#指定要消费的 Kafka 集群的主题名称
a1.sources.r1.topic = test1
#指定 Kafka 集群的消费者组 ID
a1.sources.r1.groupId = flume
```

2. Flume 的 Channel

在 Flume 的官网中可以看到表 5-2 所列的几种 Channel。

表 5-2　几种常用 Channel

Memory Channel	Event 数据存储在内存中
JDBC Channel	Event 数据存储在数据库中
File Channel	Event 数据存储在磁盘中
Spillable Memory Channel	Event 数据存储在内存和磁盘中，内存满了会溢写到磁盘中，不要在生产环境使用
Pseudo Transaction Channel	测试用途
Custom Channel	自定义 Channel
Kafka Channel	利用 Kafka 存储数据

下面了解几个主要的 Source 源：

（1） Memory Channel

Events 存储在配置最大的内存队列中。对于流量较高和由于 Agent 故障而准备丢失数据的流程来说，这是一个理想的选择。参数列表如图 5-11 所示。

Property Name	Default	Description
type	–	The component type name, needs to be memory
capacity	100	The maximum number of events stored in the channel
transactionCapacity	100	The maximum number of events the channel will take from a source or give to a sink per transaction
keep-alive	3	Timeout in seconds for adding or removing an event
byteCapacityBufferPercentage	20	Defines the percent of buffer between byteCapacity and the estimated total size of all events in the channel, to account for data in headers. See below.
byteCapacity	see description	Maximum total **bytes** of memory allowed as a sum of all events in this channel. The implementation only counts the Event body, which is the reason for providing the byteCapacityBufferPercentage configuration parameter as well. Defaults to a computed value equal to 80% of the maximum memory available to the JVM (i.e. 80% of the -Xmx value passed on the command line). Note that if you have multiple memory channels on a single JVM, and they happen to hold the same physical events (i.e. if you are using a replicating channel selector from a single source) then those event sizes may be double-counted for channel byteCapacity purposes. Setting this value to 0 will cause this value to fall back to a hard internal limit of about 200 GB.

图 5-11　Memory Channel 参数列表

具体使用方式如下：

```
#指定 Channel 的名称
```

```
2.a1.channels = c1
3.#指定 Channel 的类型
4.a1.channels.c1.type = memory
5.#指定 Channel 的容量大小
6.a1.channels.c1.capacity = 10000
7.#指定 Channel 每次事务提交的容量
8.a1.channels.c1.transactionCapacity = 10000
```

(2) JDBC Channel

Event 存储在持久化数据库中。JDBC Channel 目前支持嵌入式 Derby。这是一个持续的 Channel，对于可恢复性非常重要的流程来说是不错的选择。参数列表如图 5－12 所示。

Property Name	Default	Description
type	–	The component type name, needs to be jdbc
db.type	DERBY	Database vendor, needs to be DERBY.
driver.class	org.apache.derby.jdbc.EmbeddedDriver	Class for vendor's JDBC driver
driver.url	(constructed from other properties)	JDBC connection URL
db.username	"sa"	User id for db connection
db.password	–	password for db connection
connection.properties.file	–	JDBC Connection property file path
create.schema	true	If true, then creates db schema if not there
create.index	true	Create indexes to speed up lookups
create.foreignkey	true	
transaction.isolation	"READ_COMMITTED"	Isolation level for db session READ_UNCOMMITTED, READ_COMMITTED, SERIALIZABLE, REPEATABLE_READ
maximum.connections	10	Max connections allowed to db
maximum.capacity	0 (unlimited)	Max number of events in the channel
sysprop.*		DB Vendor specific properties
sysprop.user.home		Home path to store embedded Derby database

图 5－12　JDBC Channel 参数列表

具体使用方式如下：

```
1.#指定 Channel 的名称
2.a1.channels = c1
3.#指定 Channel 的类型
4.a1.channels.c1.type = jdbc
```

(3) File Channel

File Channel 是一个持久化的隧道，数据安全性高，并且只要磁盘空间足够，它就可以将数据存储到磁盘上。参数列表如图 5－13 所示。

具体使用方式如下：

```
1.#指定 Channel 的名称
2.a1.channels = c1
```

Property Name Default	Description	
type	–	The component type name, needs to be `file`.
checkpointDir	~/.flume/file-channel/checkpoint	The directory where checkpoint file will be stored
useDualCheckpoints	false	Backup the checkpoint. If this is set to `true`, `backupCheckpointDir` **must** be set
backupCheckpointDir	–	The directory where the checkpoint is backed up to. This directory **must not** be the same as the data directories or the checkpoint directory
dataDirs	~/.flume/file-channel/data	Comma separated list of directories for storing log files. Using multiple directories on separate disks can improve file channel peformance
transactionCapacity	10000	The maximum size of transaction supported by the channel
checkpointInterval	30000	Amount of time (in millis) between checkpoints
maxFileSize	2146435071	Max size (in bytes) of a single log file
minimumRequiredSpace	524288000	Minimum Required free space (in bytes). To avoid data corruption, File Channel stops accepting take/put requests when free space drops below this value
capacity	1000000	Maximum capacity of the channel
keep-alive	3	Amount of time (in sec) to wait for a put operation
use-log-replay-v1	false	Expert: Use old replay logic
use-fast-replay	false	Expert: Replay without using queue
checkpointOnClose	true	Controls if a checkpoint is created when the channel is closed. Creating a checkpoint on close speeds up subsequent startup of the file channel by avoiding replay.
encryption.activeKey	–	Key name used to encrypt new data
encryption.cipherProvider	–	Cipher provider type, supported types: AESCTRNOPADDING
encryption.keyProvider	–	Key provider type, supported types: JCEKSFILE
encryption.keyProvider.keyStoreFile	–	Path to the keystore file
encrpytion.keyProvider.keyStorePasswordFile	–	Path to the keystore password file
encryption.keyProvider.keys	–	List of all keys (e.g. history of the activeKey setting)
encyption.keyProvider.keys.*.passwordFile	–	Path to the optional key password file

图 5－13 File Channel 参数列表

```
#指定 Channel 的类型
a1.channels.c1.type=file
#指定检查点存储目录
a1.channels.c1.checkpointDir=/mnt/flume/checkpoint
#指定 Channel 存储数据的磁盘目录
a1.channels.c1.dataDirs=/mnt/flume/data
```

（4）Kafka Channel

Event 存储在 Kafka 集群中。Kafka 提供高可用性和高可靠性，所以，当 Agent 或者 Kafka Broker 崩溃时，Event 能马上被其他 Sink 使用。参数列表如图 5－14 所示。

具体使用方式如下：

```
#指定 Channel 的名称
a1.channels=c1
#指定 Channel 的类型
a1.channels.c1.type=org.apache.flume.channel.kafka.KafkaChannel
#指定 Channel 的容量大小
a1.channels.c1.capacity=10000
#指定 Channel 每次事务提交的容量
```

Property Name	Default	Description
type	-	The component type name, needs to be org.apache.flume.channel.kafka.KafkaChannel
brokerList	-	List of brokers in the Kafka cluster used by the channel This can be a partial list of brokers, but we recommend at least two for HA. The format is comma separated list of hostname:port
zookeeperConnect	-	URI of ZooKeeper used by Kafka cluster The format is comma separated list of hostname:port. If chroot is used, it is added once at the end. For example: zookeeper-1:2181,zookeeper-2:2182,zookeeper-3:2181/kafka
topic	flume-channel	Kafka topic which the channel will use
groupId	flume	Consumer group ID the channel uses to register with Kafka. Multiple channels must use the same topic and group to ensure that when one agent fails another can get the data Note that having non-channel consumers with the same ID can lead to data loss.
parseAsFlumeEvent	true	Expecting Avro datums with FlumeEvent schema in the channel. This should be true if Flume source is writing to the channel And false if other producers are writing into the topic that the channel is using Flume source messages to Kafka can be parsed outside of Flume by using org.apache.flume.source.avro.AvroFlumeEvent provided by the flume-ng-sdk artifact
readSmallestOffset	false	When set to true, the channel will read all data in the topic, starting from the oldest event when false, it will read only events written after the channel started When "parseAsFlumeEvent" is true, this will be false. Flume source will start prior to the sinks and this guarantees that events sent by source before sinks start will not be lost.
Other Kafka Properties	-	These properties are used to configure the Kafka Producer and Consumer used by the channel. Any property supported by Kafka can be used. The only requirement is to prepend the property name with the prefix kafka.. For example: kafka.producer.type

图 5 – 14 Kafka Channel 参数列表

```
a1.channels.c1.transactionCapacity =1000
#指定 Kafka 集群的节点名称
a1.channels.c1.brokerList =kafka -2:9092,kafka -3:9092
#指定要消费的 Kafka 集群主题的名称
a1.channels.c1.topic =channel1
#指定 Kafka 集群的 ZooKeeper 地址
a1.channels.c1.zookeeperConnect =kafka -1:2181
```

（5） Custom Channel

自定义 Channel 是自己实现 Channel 接口。当 Flume Agent 启动时，一个自定义 Channel 类和它的依赖项必须包含在 Agent 的 classpath 中。参数列表如图 5 – 15 所示。

Property Name	Default	Description
type	-	The component type name, needs to be a FQCN

图 5 – 15 Custom Channel 参数列表

具体使用方式如下：

```
#指定 Channel 的名称
a1.channels =c1
#指定 Channel 的类型(自定义 Channel 类)
a1.channels.c1.type =org.example.MyChannel
```

3. Flume 的 Sink

在 Flume 的官网中可以看到表 5－3 所列的几种 Sink。

表 5－3　几种常见 Sink

HDFS Sink	数据写入 HDFS
Logger Sink	数据写入日志文件
Avro Sink	数据被转换为 Avro Event，然后发送到配置的 RPC 端口
Thrift Sink	数据被转换为 Thrift Event，然后发送到配置的 RPC 端口
IRC Sink	数据在 IRC 上进行回放
File Roll Sink	存储数据到本地文件系统
Null Sink	丢弃所有数据
HBase Sink	数据写入 HBase 数据库
Hive Sink	数据写入 Hive 数据仓库
MorphlineSolr Sink	数据发送到 Solr 服务器
Elastic Search Sink	数据发送到 ES 服务器
Kafka Sink	数据发送到 Kafka 集群
Custom Sink	用户自定义 Sink

下面了解几个主要的 Sink。

（1） HDFS Sink

该 Sink 把 Event 写进 Hadoop 分布式文件系统（HDFS）。目前它支持创建文本和序列文件，支持在两种文件类型间压缩。文件可以基于数据的经过时间或大小或事件的数量而周期性地滚动，还可以通过属性（如时间戳或发生事件的机器）把数据划分为桶或区。参数列表如图 5－16 所示。

具体使用方式如下：

```
#指定 Channel 的名称
a1.channels = c1
#指定 Sink 的名称
a1.sinks = k1
#指定 Sink 的类型
a1.sinks.k1.type = hdfs
#将 Sink 跟 Channel 连接在一起
a1.sinks.k1.channel = c1
#指定 HDFS 存储数据的目录(% 表示会读取当前系统的时间,按照时间生成目录)
a1.sinks.k1.hdfs.path = /flume/events/% y - % m - % d/% H% M/% S
```

Name	Default	Description
channel	–	
type	–	The component type name, needs to be `hdfs`
hdfs.path	–	HDFS directory path (eg hdfs://namenode/flume/webdata/)
hdfs.filePrefix	FlumeData	Name prefixed to files created by Flume in hdfs directory
hdfs.fileSuffix	–	Suffix to append to file (eg `.avro` - *NOTE: period is not automatically added*)
hdfs.inUsePrefix	–	Prefix that is used for temporal files that flume actively writes into
hdfs.inUseSuffix	`.tmp`	Suffix that is used for temporal files that flume actively writes into
hdfs.rollInterval	30	Number of seconds to wait before rolling current file (0 = never roll based on time interval)
hdfs.rollSize	1024	File size to trigger roll, in bytes (0: never roll based on file size)
hdfs.rollCount	10	Number of events written to file before it rolled (0 = never roll based on number of events)
hdfs.idleTimeout	0	Timeout after which inactive files get closed (0 = disable automatic closing of idle files)
hdfs.batchSize	100	number of events written to file before it is flushed to HDFS
hdfs.codeC	–	Compression codec. one of following : gzip, bzip2, lzo, lzop, snappy
hdfs.fileType	SequenceFile	File format: currently `SequenceFile`, `DataStream` or `CompressedStream` (1)DataStream will not compress output file and please don't set codeC (2)CompressedStream requires set hdfs.codeC with an available codeC
hdfs.maxOpenFiles	5000	Allow only this number of open files. If this number is exceeded, the oldest file is closed.
hdfs.minBlockReplicas	–	Specify minimum number of replicas per HDFS block. If not specified, it comes from the default Hadoop config in the classpath.
hdfs.writeFormat	–	Format for sequence file records. One of "Text" or "Writable" (the default).
hdfs.callTimeout	10000	Number of milliseconds allowed for HDFS operations, such as open, write, flush, close. This number should be increased if many HDFS timeout operations are occurring.
hdfs.threadsPoolSize	10	Number of threads per HDFS sink for HDFS IO ops (open, write, etc.)
hdfs.rollTimerPoolSize	1	Number of threads per HDFS sink for scheduling timed file rolling
hdfs.kerberosPrincipal	–	Kerberos user principal for accessing secure HDFS
hdfs.kerberosKeytab	–	Kerberos keytab for accessing secure HDFS
hdfs.proxyUser		
hdfs.round	false	Should the timestamp be rounded down (if true, affects all time based escape sequences except %t)
hdfs.roundValue	1	Rounded down to the highest multiple of this (in the unit configured using `hdfs.roundUnit`), less than current time.
hdfs.roundUnit	second	The unit of the round down value - `second`, `minute` or `hour`.
hdfs.timeZone	Local Time	Name of the timezone that should be used for resolving the directory path, e.g. America/Los_Angeles.
hdfs.useLocalTimeStamp	false	Use the local time (instead of the timestamp from the event header) while replacing the escape sequences.
hdfs.closeTries	0	Number of times the sink must try renaming a file, after initiating a close attempt. If set to 1, this sink will not re-try a failed rename (due to, for example, NameNode or DataNode failure), and may leave the file in an open state with a .tmp extension. If set to 0, the sink will try to rename the file until the file is eventually renamed (there is no limit on the number of times it would try). The file may still remain open if the close call fails but the data will be intact and in this case, the file will be closed only after a Flume restart.
hdfs.retryInterval	180	Time in seconds between consecutive attempts to close a file. Each close call costs multiple RPC round-trips to the Namenode, so setting this too low can cause a lot of load on the name node. If set to 0 or less, the sink will not attempt to close the file if the first attempt fails, and may leave the file open or with a ".tmp" extension.

图 5－16 HDFS Sink 参数列表

```
#指定存储数据文件的前缀
a1.sinks.k1.hdfs.filePrefix = events -
#是否使用时间戳来生成文件夹
a1.sinks.k1.hdfs.round = true
#使用时间戳来生成文件夹的间隔值
a1.sinks.k1.hdfs.roundValue = 10
#使用时间戳来生成文件夹的单位名称
a1.sinks.k1.hdfs.roundUnit = minute
```

(2) Logger Sink

使用日志的方式存储数据，Logs Event 在 INFO 水平。典型用法是测试或者调试。参数列表如图 5－17 所示。

具体使用方式如下：

```
#指定 Channel 的名称
a1.channels = c1
#指定 Sink 的名称
```

Property Name	Default	Description
channel	–	
type	–	The component type name, needs to be `logger`
maxBytesToLog	16	Maximum number of bytes of the Event body to log

图 5 – 17　Logger Sink 参数列表

```
a1.sinks = k1
#指定 Sink 的类型
a1.sinks.k1.type = logger
#将 Sink 跟 Channel 连接在一起
a1.sinks.k1.channel = c1
```

（3） Avro Sink

将 Flume Event 发送到 Sink，转换为 Avro Event，并发送到配置好的 hostname/port。从配置好的 Channel 中按照大小批量获取 Event。参数列表如图 5 – 18 所示。

Property Name	Default	Description
channel	–	
type	–	The component type name, needs to be `avro`.
hostname	–	The hostname or IP address to bind to.
port	–	The port # to listen on.
batch-size	100	number of event to batch together for send.
connect-timeout	20000	Amount of time (ms) to allow for the first (handshake) request.
request-timeout	20000	Amount of time (ms) to allow for requests after the first.
reset-connection-interval	none	Amount of time (s) before the connection to the next hop is reset. This will force the Avro Sink to reconnect to the next hop. This will allow the sink to connect to hosts behind a hardware load-balancer when news hosts are added without having to restart the agent.
compression-type	none	This can be "none" or "deflate". The compression-type must match the compression-type of matching AvroSource
compression-level	6	The level of compression to compress event. 0 = no compression and 1-9 is compression. The higher the number the more compression
ssl	false	Set to true to enable SSL for this AvroSink. When configuring SSL, you can optionally set a "truststore", "truststore-password", "truststore-type", and specify whether to "trust-all-certs".
trust-all-certs	false	If this is set to true, SSL server certificates for remote servers (Avro Sources) will not be checked. This should NOT be used in production because it makes it easier for an attacker to execute a man-in-the-middle attack and "listen in" on the encrypted connection.
truststore	–	The path to a custom Java truststore file. Flume uses the certificate authority information in this file to determine whether the remote Avro Source's SSL authentication credentials should be trusted. If not specified, the default Java JSSE certificate authority files (typically "jssecacerts" or "cacerts" in the Oracle JRE) will be used.
truststore-password	–	The password for the specified truststore.
truststore-type	JKS	The type of the Java truststore. This can be "JKS" or other supported Java truststore type.
exclude-protocols	SSLv2Hello SSLv3	Space-separated list of SSL/TLS protocols to exclude
maxIoWorkers	2 * the number of available processors in the machine	The maximum number of I/O worker threads. This is configured on the NettyAvroRpcClient NioClientSocketChannelFactory.

图 5 – 18　Avro Sink 参数列表

具体使用方式如下：

```
#指定 Channel 的名称
a1.channels = c1
#指定 Sink 的名称
a1.sinks = k1
#指定 Sink 的类型
```

```
6.a1.sinks.k1.type=avro
7.#将 Sink 跟 Channel 连接在一起
8.a1.sinks.k1.channel=c1
9.#Avro 接收数据的节点名称
10.a1.sinks.k1.hostname=10.10.10.10
11.#Avro 接收数据节点的端口号
12.a1.sinks.k1.port=4545
```

(4) Thrift Sink

将 Flume Event 发送到 Sink，转换为 Thrift Event，并发送到配置好的 hostname/port。从配置好的 Channel 中按照大小批量获取 Event。参数列表如图 5－19 所示。

Property Name	Default	Description
channel	–	
type	–	The component type name, needs to be thrift.
hostname	–	The hostname or IP address to bind to.
port	–	The port # to listen on.
batch-size	100	number of event to batch together for send.
connect-timeout	20000	Amount of time (ms) to allow for the first (handshake) request.
request-timeout	20000	Amount of time (ms) to allow for requests after the first.
connection-reset-interval	none	Amount of time (s) before the connection to the next hop is reset. This will force the Thrift Sink to reconnect to the next hop. This will allow the sink to connect to hosts behind a hardware load-balancer when news hosts are added without having to restart the agent.
ssl	false	Set to true to enable SSL for this ThriftSink. When configuring SSL, you can optionally set a "truststore", "truststore-password" and "truststore-type"
truststore	–	The path to a custom Java truststore file. Flume uses the certificate authority information in this file to determine whether the remote Thrift Source's SSL authentication credentials should be trusted. If not specified, the default Java JSSE certificate authority files (typically "jssecacerts" or "cacerts" in the Oracle JRE) will be used.
truststore-password	–	The password for the specified truststore.
truststore-type	JKS	The type of the Java truststore. This can be "JKS" or other supported Java truststore type.
exclude-protocols	SSLv3	Space-separated list of SSL/TLS protocols to exclude
kerberos	false	Set to true to enable kerberos authentication. In kerberos mode, client-principal, client-keytab and server-principal are required for successful authentication and communication to a kerberos enabled Thrift Source.
client-principal	—	The kerberos principal used by the Thrift Sink to authenticate to the kerberos KDC.
client-keytab	—	The keytab location used by the Thrift Sink in combination with the client-principal to authenticate to the kerberos KDC.
server-principal	–	The kerberos principal of the Thrift Source to which the Thrift Sink is configured to connect to.

图 5－19　Thrift Sink 参数列表

具体使用方式如下：

```
1.#指定 Channel 的名称
2.a1.channels=c1
3.#指定 Sink 的名称
4.a1.sinks=k1
5.#指定 Sink 的类型
6.a1.sinks.k1.type=thrift
7.#将 Sink 跟 Channel 连接在一起
8.a1.sinks.k1.channel=c1
9.#Avro 接收数据的节点名称
```

```
10.a1.sinks.k1.hostname =10.10.10.10
11.#Avro 接收数据节点的端口号
12.a1.sinks.k1.port =4545
```

（5）File Roll Sink

本地文件系统存储 Event。参数列表如图 5－20 所示。

Property Name	Default	Description
channel	-	
type	-	The component type name, needs to be file_roll.
sink.directory	-	The directory where files will be stored
sink.rollInterval	30	Roll the file every 30 seconds. Specifying 0 will disable rolling and cause all events to be written to a single file.
sink.serializer	TEXT	Other possible options include avro_event or the FQCN of an implementation of EventSerializer.Builder interface.
batchSize	100	

图 5－20　File Roll Sink 参数列表

具体使用方式如下：

```
1.#指定 Channel 的名称
2.a1.channels = c1
3.#指定 Sink 的名称
4.a1.sinks = k1
5.#指定 Sink 的类型
6.a1.sinks.k1.type = file_roll
7.#将 Sink 跟 Channel 连接在一起
8.a1.sinks.k1.channel = c1
9.#指定存储数据的目录
10.a1.sinks.k1.sink.directory = /var/log/flume
```

（6）HBase Sink

该 Sink 写数据到 HBase。参数列表如图 5－21 所示。

Property Name	Default	Description
channel	-	
type	-	The component type name, needs to be hbase
table	-	The name of the table in Hbase to write to.
columnFamily	-	The column family in Hbase to write to.
zookeeperQuorum	-	The quorum spec. This is the value for the property hbase.zookeeper.quorum in hbase-site.xml
znodeParent	/hbase	The base path for the znode for the -ROOT- region. Value of zookeeper.znode.parent in hbase-site.xml
batchSize	100	Number of events to be written per txn.
coalesceIncrements	false	Should the sink coalesce multiple increments to a cell per batch. This might give better performance if there are multiple increments to a limited number of cells.
serializer	org.apache.flume.sink.hbase.SimpleHbaseEventSerializer	Default increment column = "iCol", payload column = "pCol".
serializer.*	-	Properties to be passed to the serializer.
kerberosPrincipal	-	Kerberos user principal for accessing secure HBase
kerberosKeytab	-	Kerberos keytab for accessing secure HBase

图 5－21　HBase Sink 参数列表

具体使用方式如下：

```
1.#指定 Channel 的名称
2.a1.channels = c1
3.#指定 Sink 的名称
4.a1.sinks = k1
5.#指定 Sink 的类型
6.a1.sinks.k1.type = hbase
7.#将 Sink 跟 Channel 连接在一起
8.a1.sinks.k1.channel = c1
9.#指定 HBase 表的名称
10.a1.sinks.k1.table = foo_table
11.#指定 HBase 表的每一列的名称
12.a1.sinks.k1.columnFamily = bar_cf
13.#指定 HBase 存储的序列化类
14. a1.sinks.k1.serializer = org.apache.flume.sink.hbase.RegexHbase
  EventSerializer
```

（7）Hive Sink

该 Sink Stream 将包含分割文本或 JSON 数据的 Event 直接传送到 Hive 表或分区中，用 Hive 事务写 Event。当一系列 Event 提交到 Hive 时，它们可以马上被 Hive 查询到。参数列表如图 5－22 所示。

Name	Default	Description
channel	–	
type	–	The component type name, needs to be `hive`
hive.metastore	–	Hive metastore URI (eg thrift://a.b.com:9083)
hive.database	–	Hive database name
hive.table	–	Hive table name
hive.partition	–	Comma separate list of partition values identifying the partition to write to. May contain escape sequences. E.g: If the table is partitioned by (continent: string, country :string, time : string) then 'Asia,India,2014-02-26-01-21' will indicate continent=Asia,country=India,time=2014-02-26-01-21
hive.txnsPerBatchAsk	100	Hive grants a *batch of transactions* instead of single transactions to streaming clients like Flume. This setting configures the number of desired transactions per Transaction Batch. Data from all transactions in a single batch end up in a single file. Flume will write a maximum of batchSize events in each transaction in the batch. This setting in conjunction with batchSize provides control over the size of each file. Note that eventually Hive will transparently compact these files into larger files.
heartBeatInterval	240	(In seconds) Interval between consecutive heartbeats sent to Hive to keep unused transactions from expiring. Set this value to 0 to disable heartbeats.
autoCreatePartitions	true	Flume will automatically create the necessary Hive partitions to stream to
batchSize	15000	Max number of events written to Hive in a single Hive transaction
maxOpenConnections	500	Allow only this number of open connections. If this number is exceeded, the least recently used connection is closed.
callTimeout	10000	(In milliseconds) Timeout for Hive & HDFS I/O operations, such as openTxn, write, commit, abort.
serializer		Serializer is responsible for parsing out field from the event and mapping them to columns in the hive table. Choice of serializer depends upon the format of the data in the event. Supported serializers: DELIMITED and JSON
roundUnit	minute	The unit of the round down value - `second`, `minute` or `hour`.
roundValue	1	Rounded down to the highest multiple of this (in the unit configured using hive.roundUnit), less than current time
timeZone	Local Time	Name of the timezone that should be used for resolving the escape sequences in partition, e.g. America/Los_Angeles.
useLocalTimeStamp	false	Use the local time (instead of the timestamp from the event header) while replacing the escape sequences.

图 5－22　Hive Sink 参数列表

具体使用方式如下：

```
#指定 Channel 的名称
a1.channels = c1
#指定 Sink 的名称
a1.sinks = k1
#指定 Sink 的类型
a1.sinks.k1.type = hive
#将 Sink 与 Channel 连接在一起
a1.sinks.k1.channel = c1
#指定 Hive 的元数据地址
a1.sinks.k1.hive.metastore = thrift://127.0.0.1:9083
#指定 Hive 的数据库名称
a1.sinks.k1.hive.database = logsdb
#指定存储数据的 Hive 表名
a1.sinks.k1.hive.table = weblogs
#指定 Hive 的分区信息
a1.sinks.k1.hive.partition = asia,%{country},%y-%m-%d-%H-%M
#指定使用本地的时间戳
a1.sinks.k1.useLocalTimeStamp = false
#指定 Hive 存储数据目录的生成规则,是否使用本地时间戳
a1.sinks.k1.round = true
#使用时间戳来生成文件夹的间隔值
a1.sinks.k1.roundValue = 10
#使用时间戳来生成文件夹的单位名称
a1.sinks.k1.roundUnit = minute
#指定 Hive 表的序列化规则
a1.sinks.k1.serializer = DELIMITED
#指定 Hive 表序列化的分割字符
a1.sinks.k1.serializer.delimiter = "\t"
#指定 Hive 表的字段名称
a1.sinks.k1.serializer.fieldnames = id,msg
```

（8） MorphlineSolr Sink

该 Sink 从 Flume Event 提取数据并转换，在 Apache Solr 服务端实时加载，Apache Solr 服务端为最终用户或者搜索应用程序提供查询服务。参数列表如图 5 - 23 所示。

具体使用方式如下：

Property Name	Default	Description
channel	–	
type	–	The component type name, needs to be org.apache.flume.sink.solr.morphline.MorphlineSolrSink
morphlineFile	–	The relative or absolute path on the local file system to the morphline configuration file. Example: /etc/flume-ng/conf/morphline.conf
morphlineId	null	Optional name used to identify a morphline if there are multiple morphlines in a morphline config file
batchSize	1000	The maximum number of events to take per flume transaction.
batchDurationMillis	1000	The maximum duration per flume transaction (ms). The transaction commits after this duration or when batchSize is exceeded, whichever comes first.
handlerClass	org.apache.flume.sink.solr.morphline.MorphlineHandlerImpl	The FQCN of a class implementing org.apache.flume.sink.solr.morphline.MorphlineHandler

图 5－23 MorphlineSolr Sink 参数列表

```
#指定 Channel 的名称
a1.channels = c1
#指定 Sink 的名称
a1.sinks = k1
#指定 Sink 的类型
a1.sinks.k1.type = org.apache.flume.sink.solr.morphline.MorphlineSolrSink
#将 Sink 与 Channel 连接在一起
a1.sinks.k1.channel = c1
#指定文件存储的目录
a1.sinks.k1.morphlineFile = /etc/flume-ng/conf/morphline.conf
```

(9) Elastic Search Sink

该 Sink 写数据到 Elastic Search 集群。参数列表如图 5－24 所示。

具体使用方式如下：

```
#指定 Channel 的名称
a1.channels = c1
#指定 Sink 的名称
a1.sinks = k1
#指定 Sink 的类型
a1.sinks.k1.type = elasticsearchsink
#将 Sink 与 Channel 连接在一起
a1.sinks.k1.channel = c1
#设置 ES 集群的节点名称
a1.sinks.k1.hostNames = 127.0.0.1:9200,127.0.0.2:9300
#设置 ES 集群的索引名称
```

Property Name	Default	Description
channel	–	
type	–	The component type name, needs to be org.apache.flume.sink.elasticsearch.ElasticSearchSink
hostNames	–	Comma separated list of hostname:port, if the port is not present the default port '9300' will be used
indexName	flume	The name of the index which the date will be appended to. Example 'flume' -> 'flume-yyyy-MM-dd' Arbitrary header substitution is supported, eg. %{header} replaces with value of named event header
indexType	logs	The type to index the document to, defaults to 'log' Arbitrary header substitution is supported, eg. %{header} replaces with value of named event header
clusterName	elasticsearch	Name of the ElasticSearch cluster to connect to
batchSize	100	Number of events to be written per txn.
ttl	–	TTL in days, when set will cause the expired documents to be deleted automatically, if not set documents will never be automatically deleted. TTL is accepted both in the earlier form of integer only e.g. a1.sinks.k1.ttl = 5 and also with a qualifier ms (millisecond), s (second), m (minute), h (hour), d (day) and w (week). Example a1.sinks.k1.ttl = 5d will set TTL to 5 days. Follow http://www.elasticsearch.org/guide/reference/mapping/ttl-field/ for more information.
serializer	org.apache.flume.sink.elasticsearch.ElasticSearchLogStashEventSerializer	The ElasticSearchIndexRequestBuilderFactory or ElasticSearchEventSerializer to use. Implementations of either class are accepted but ElasticSearchIndexRequestBuilderFactory is preferred.
serializer.*	–	Properties to be passed to the serializer.

图 5 –24　Elastic Search Sink 参数列表

```
a1.sinks.k1.indexName = foo_index
#设置 ES 集群的索引类型
a1.sinks.k1.indexType = bar_type
#设置 ES 集群的集群名称
a1.sinks.k1.clusterName = foobar_cluster
#设置每次写入 ES 集群的记录条数
a1.sinks.k1.batchSize = 500
#设置 ES 集群的数据存活时间
a1.sinks.k1.ttl = 5d
#设置 ES 集群的序列化类
a1.sinks.k1.serializer = org.apache.flume.sink.elasticsearch.
  ElasticSearchDynamicSerializer
```

（10）Kafka Sink

该 Sink 可以实现导出数据到一个 Kafka Topic。参数列表如图 5 –25 所示。

具体使用方式如下：

```
#指定 Channel 的名称
a1.channels = c1
#指定 Sink 的名称
```

Property Name	Default	Description
type	-	Must be set to `org.apache.flume.sink.kafka.KafkaSink`
brokerList	-	List of brokers Kafka-Sink will connect to, to get the list of topic partitions This can be a partial list of brokers, but we recommend at least two for HA. The format is comma separated list of hostname:port
topic	default-flume-topic	The topic in Kafka to which the messages will be published. If this parameter is configured, messages will be published to this topic. If the event header contains a "topic" field, the event will be published to that topic overriding the topic configured here.
batchSize	100	How many messages to process in one batch. Larger batches improve throughput while adding latency.
requiredAcks	1	How many replicas must acknowledge a message before its considered successfully written. Accepted values are 0 (Never wait for acknowledgement), 1 (wait for leader only), -1 (wait for all replicas) Set this to -1 to avoid data loss in some cases of leader failure.
Other Kafka Producer Properties	-	These properties are used to configure the Kafka Producer. Any producer property supported by Kafka can be used. The only requirement is to prepend the property name with the prefix `kafka.`. For example: kafka.producer.type

图5－25　Kafka Sink 参数列表

```
a1.sinks = k1
#指定 Sink 的类型
a1.sinks.k1.type = org.apache.flume.sink.kafka.KafkaSink
#将 Sink 与 Channel 连接在一起
a1.sinks.k1.channel = c1
#指定写入 Kafka 集群的主题名称
a1.sinks.k1.topic = mytopic
#指定 Kafka 集群节点的名称
a1.sinks.k1.brokerList = localhost:9092
#指定每次写入 Kafka 集群的数据条数
a1.sinks.k1.batchSize = 20
```

（11） Custom Sink

自定义 Sink 是实现 Sink 接口。当启动 Flume Agent 时，一个自定义 Sink 类和它的依赖项必须在 Agent 的 classpath 中。参数列表如图 5－26 所示。

Property Name	Default	Description
channel	-	
type	-	The component type name, needs to be your FQCN

图5－26　Custom Sink 参数列表

具体使用方式如下：

```
#指定 Channel 的名称
a1.channels = c1
#指定 Sink 的名称
a1.sinks = k1
```

```
5.#指定 Sink 的类型
6.a1.sinks.k1.type=org.example.MySink
7.#将 Sink 与 Channel 连接在一起
8.a1.sinks.k1.channel=c1
```

4. Flume 底层运行机制

每个 Flume Agent 包含三个主要组件：Source、Channel、Sink，这里将学习这些组件及它们是如何协同工作的。

Source 是从一些其他产生数据的应用中接收数据的活跃组件，有的可以自己产生数据，但是这样的 Source 通常是用于测试的。Source 可以监听一个或多个网络端口，用于接收数据或从本地文件系统读取数据。每个 Source 必须至少连接一个 Channel。基于一些标准，一个 Source 可以写入几个 Channel，复制事件到所有或某些 Channel 中。

Channel 的行为与队列的类似，Source 写入它们，Sink 从它们中读取。多个 Source 可以安全地写入到相同的 Channel，并且多个 Sink 可以从相同的 Channel 中读取，但是一个 Sink 只能从一个 Channel 中读取。假设多个 Sink 从相同的 Channel 中读取，则只有一个 Sink 可以从 Channel 中读取一个指定的事件。

Sink 连续轮询各自的 Channel 来读取和删除事件。Sink 将事件推动到下一阶段或者最终目的地。一旦在下一阶段或者目的地中数据是安全的，Sink 就通过事务提交来通知 Channel，可以从 Channel 中删除这些事件。图 5－27 是对 Flume Agent 流程的一个简单描述。

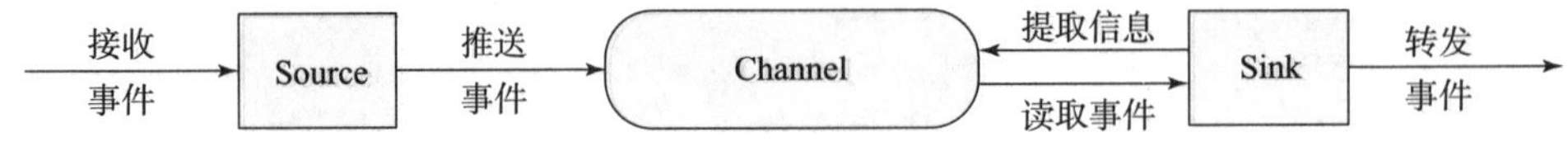

图 5－27　Flume Agent 流程

Flume 本身不限制 Agent 的 Source、Channel 和 Sink 的数量，因此，Flume Source 可以接收事件，并能通过配置将事件复制到多个目的地，这使得 Source 通过 Channel 处理器、拦截器和 Channel 选择器，将数据写入 Channel 成为可能。

每个 Source 都拥有自己的 Channel 处理器。每次 Source 将数据写入 Channel，是通过委派任务形式到其 Channel 处理器来完成的。然后 Channel 处理器将这些事件传到一个或多个由 Source 配置的拦截器中。

如果写入必需的 Channel 失败，会导致 Channel 处理器抛出 ChannelException 异常，表明 Source 必须重试该事件（实际上，所有的事件都在所属事务中）。一旦写出事件，处理器会发送确认信息给发送该事件的系统，并继续接收更多的事件。图 5－28 所示为 Channel 运行流程。

Sink 运行器（Sink Runner）在一个 Sink 组（Sink Group）里运行，Sink 组包含有一个或者多个 Sink。如果组中只存在一个 Sink，则没有组将更有效率。Sink 运行器仅仅是一个通过询问 Sink 组来处理下一批事件的线程。每个 Sink 组有一个 Sink 处理器，处理器选择组中的 Sink 去处理下一个事件集合。尽管多个 Sink 可以从同一个 Channel 获取数据，但是每个

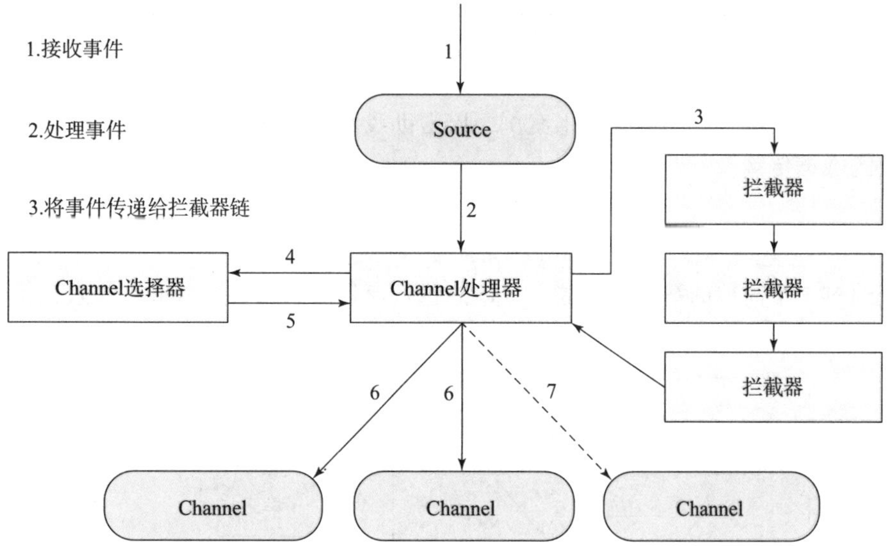

图 5 - 28　Channel 运行流程

Sink 只能从一个 Channel 中获取数据。选定的 Sink 从 Channel 中接收事件，并将事件写入下一阶段的最终目的地。流程如图 5 - 29 所示。

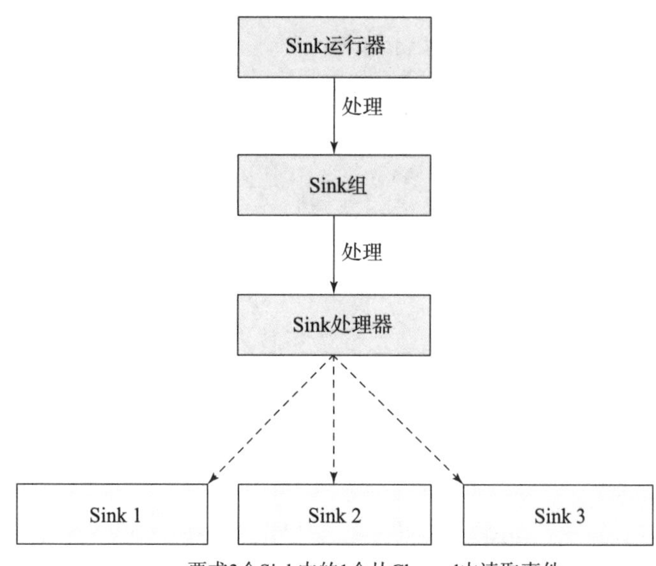

图 5 - 29　Sink 运行流程

任务 3　Flume 应用案例

本任务将介绍运行 Flume 的应用案例，以帮助我们在做项目时选择合理的架构来实现快捷、高效的数据传输。

1. 场景一

首先通过一个简单的案例来描述 Flume 的执行流程。数据通过 Netcat 输入，将结果输出到 Flume 控制台日志中。流程图如图 5－30 所示。

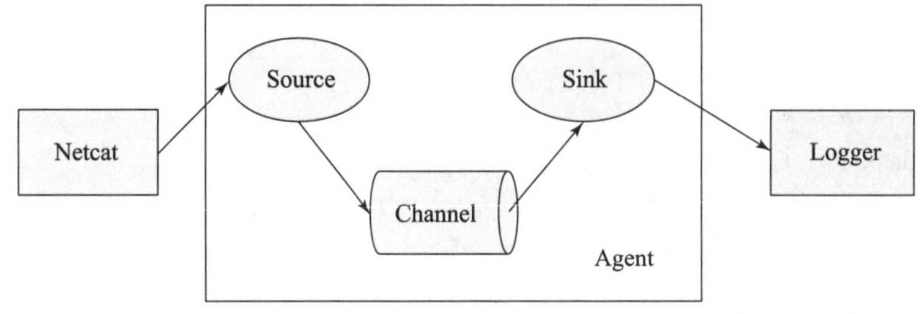

图 5－30　单台 Flume 执行流程

场景一：架构单台 FLUME 进行数据采集

操作步骤如下：

①选择安装 Flume 的节点 Superintendent2，在 root 目录下创建一个新的文件夹 Flume。

```
#创建 Flume 文件夹
[root@Superintendent2 ~]# mkdir flume
```

②进入 Flume 文件夹，创建新的文件 option。

```
#进入 Flume 文件夹
[root@Superintendent2 ~]# cd flume/
#新建 option 文件
[root@Superintendent2 flume]# vi option
```

③将如下配置信息添加到 option 文件下。

```
# 单节点的 Flume 配置
# 配置三个组件的名称
# 配置 Source 的名称
a1.sources = r1
# 配置 Sink 的名称
a1.sinks = k1
# 配置 Channel 的名称
a1.channels = c1
# 配置 Source 的类型
```

```
a1.sources.r1.type = netcat
# 配置接收 Natcat 数据的节点的名称
a1.sources.r1.bind = Superintendent2
# 配置接收 Natcat 数据的节点的端口
a1.sources.r1.port = 44444
# 描述 Sink 的类型为 Logger
a1.sinks.k1.type = logger
# 使用基于内存的 Channel 来缓存数据
a1.channels.c1.type = memory
# 配置 Channel 的容量大小
a1.channels.c1.capacity = 1000
# 配置 Channel 每次提交数据的容量
a1.channels.c1.transactionCapacity = 100
# 将 Source 和 Channel 连接在一起
a1.sources.r1.channels = c1
# 将 Source 和 Sink 连接在一起
a1.sinks.k1.channel = c1
```

④运行 Flume 启动执行。

```
#flume-ng 是 flume 的启动命令，--conf-file 指定配置文件，--name 是 Flume 的 Agent 的名称，-Dflume.root.logger 设置日志级别及显示位置
flume-ng agent --conf-file example.conf --name a1 -Dflume.root.logger=INFO,console
```

⑤Flume 启动完成之后，结果如下：

```
20/04/03 19:41:23 INFO node.PollingPropertiesFileConfigurationProvider:Configuration provider starting
20/04/03 19:41:23 INFO node.PollingPropertiesFileConfigurationProvider:Reloading configuration file:option
20/04/03 19:41:23 INFO conf.FlumeConfiguration:Added sinks:k1 Agent:a1
20/04/03 19:41:23 INFO conf.FlumeConfiguration:Processing:k1
20/04/03 19:41:23 INFO conf.FlumeConfiguration:Processing:k1
20/04/03 19:41:23 INFO conf.FlumeConfiguration: Post-validation flume configuration contains configuration for agents:[a1]
20/04/03 19:41:23 INFO node.AbstractConfigurationProvider:Creating channels
```

```
20/04/03 19:41:23 INFO channel.DefaultChannelFactory:Creating instance of channel c1 type memory
20/04/03 19:41:23 INFO node.AbstractConfigurationProvider:Created channel c1
20/04/03 19:41:23 INFO source.DefaultSourceFactory:Creating instance of source r1,type netcat
20/04/03 19:41:23 INFO sink.DefaultSinkFactory:Creating instance of sink:k1,type:logger
20/04/03 19:41:23 INFO node.AbstractConfigurationProvider:Channel c1 connected to[r1,k1]
20/04/03 19:41:23 INFO node.Application:Starting new configuration:{ sourceRunners:{r1=EventDrivenSourceRunner:{ source:org.apache.f
lume.source.NetcatSource{name:r1,state:IDLE} }} sinkRunners:{k1=SinkRunner:{ policy:org.apache.flume.sink.DefaultSinkProcessor@f36a955 counterGroup:{ name:null counters:{} } }} channels:{c1=org.apache.flume.channel.MemoryChannel{name:c1}} }20/04/03 19:41:23 INFO node.Application:Starting Channel c1
20/04/03 19:41:23 INFO instrumentation.MonitoredCounterGroup:Monitored counter group for type:CHANNEL,name:c1:Successfully regist
ered new MBean.20/04/03 19:41:23 INFO instrumentation.MonitoredCounterGroup:Component type:CHANNEL,name:c1 started
20/04/03 19:41:23 INFO node.Application:Starting Sink k1
20/04/03 19:41:23 INFO node.Application:Starting Source r1
20/04/03 19:41:23 INFO source.NetcatSource:Source starting
20/04/03 19:41:23 INFO source.NetcatSource:Created serverSocket:sun.nio.ch.ServerSocketChannelImpl[/192.168.3.192:44444]
```

⑥选择 Superintendent1 节点安装 Telnet，命令如下：

```
[root@Superintendent1 ~]# yum install telnet -y
Loaded plugins:fastestmirror
Loading mirror speeds from cached hostfile
* base:mirrors.aliyun.com
* extras:mirrors.aliyun.com
* updates:mirrors.aliyun.com
```

```
Setting up Install Process
Resolving Dependencies
-->Running transaction check
--->Package telnet.x86_64 1:0.17 -48.el6 will be installed
-->Finished Dependency Resolution
Dependencies Resolved
=============================================================================
Package Arch Version Repository Size
=============================================================================
Installing:
telnet x86_64 1:0.17 -48.el6 base 58 k
Transaction Summary
=============================================================================
Install 1 Package(s)
Total download size:58 k
Installed size:109 k
Downloading Packages:
telnet -0.17 -48.el6.x86_64.rpm |58 kB 00:00
Running rpm_check_debug
Running Transaction Test
Transaction Test Succeeded
Running Transaction
Installing:1:telnet -0.17 -48.el6.x86_64 1/1
Verifying:1:telnet -0.17 -48.el6.x86_64 1/1
Installed:
telnet.x86_64 1:0.17 -48.el6
Complete!
```

⑦安装完成之后运行 Telnet，命令如下：

```
[root@Superintendent1 ~]# telnet Superintendent2 44444
Trying 192.168.3.192...
Connected to Superintendent2.
Escape character is '^]'.
```

⑧在 Telnet 的命令行中输入以下字符串：

```
[root@Superintendent1 ~]# telnet Superintendent2 44444
Trying 192.168.3.192...
```

```
Connected to Superintendent2.
Escape character is '^]'.
#输入字符串 Flume
flume
#表示输入成功
OK
hello
OK
```

⑨查看 Flume 的显示状况：

```
20/04/03 19:41:23 INFO source.NetcatSource:Created serverSock-
  et:sun.nio.ch.ServerSocketChannelImpl[/192.168.3.192:44444]
20/04/03 19:44:34 INFO sink.LoggerSink:Event:{ headers:{} body:
  66 6C 75 6D 65 0D flume. }
20/04/03 19:44:35 INFO sink.LoggerSink:Event:{ headers:{} body:
  68 65 6C 6C 6F 0D hello. }
```

由运行结果可见，在 Telnet 中输入的“flume”和“hello”在 Flume 中显现出来，表示 Flume 的执行流程中，数据通过 Netcat 输入，并将结果输出到 Flume 控制台日志中运行成功。

场景二：架构两台 FLUME 进行数据采集

2. 场景二

在实际的生产环境中，很少出现单台 Flume 使用的情况，更多的是多台 Flume 组合使用。两台 Flume 组合使用的架构如图 5－31 所示。

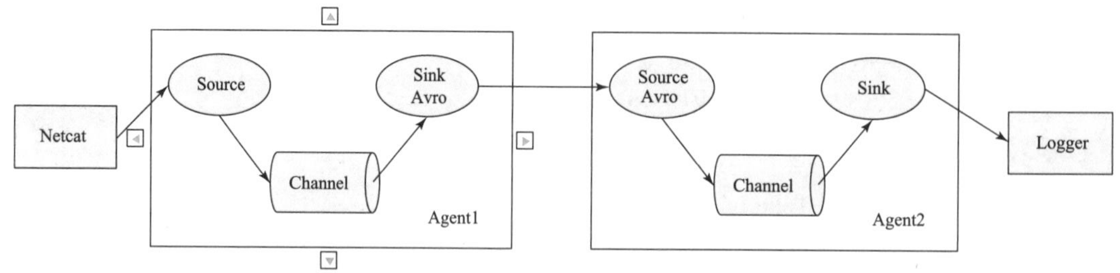

图 5－31　两台 Flume 组合使用的架构

两台 Flume Agent，Agent1 从 Netcat 接收数据，通过 Avro 将数据发送到 Agent2，Agent2 通过 Avro 接收数据，并通过 Logger 进行日志的展示。

操作步骤如下：

①在 Superintendent1 上安装 Flume，将 Superintendent1 作为 Agent1，将 Superintendent2 作为 Agent2。

②在 Superintendent1 的 root 目录下创建 Flume 文件夹，在 Flume 文件夹下创建 option_ agent1 文件，并添加配置。具体步骤如下：

```
#创建目录
[root@Superintendent1 ~]# mkdir flume
#跳转到 flume 目录
[root@Superintendent1 ~]# cd flume/
#创建 option_agent1 文件
[root@Superintendent1 flume]# vi option_agent1
#添加以下配置信息
#配置 Source 的名称
a1.sources = r1
#配置 Sink 的名称
a1.sinks = k1
#配置 Channel 的名称
a1.channels = c1
#配置 Source 的类型是 Netcat
a1.sources.r1.type = netcat
#配置接收 Source 数据的节点的名称
a1.sources.r1.bind = Superintendent1
#配置接收 Source 数据的节点的端口
a1.sources.r1.port = 44444
#配置 Sink 的类型为 Avro
a1.sinks.k1.type = avro
#配置接收 Sink 数据的节点的名称
a1.sinks.k1.hostname = Superintendent2
#配置接收 Sink 数据的节点的端口
a1.sinks.k1.port = 60000
#配置缓存数据的 Channel 类型为内存
a1.channels.c1.type = memory
#配置 Channel 的容量
a1.channels.c1.capacity = 1000
#配置 Channel 每次提交事务的记录条数
a1.channels.c1.transactionCapacity = 100
#将 Source 和 Channel 连接起来
a1.sources.r1.channels = c1
#将 Sink 和 Channel 连接起来
a1.sinks.k1.channel = c1
```

③在 Superintendent2 的/root/flume 文件夹下创建 option_agent2 文件，并添加如下配置，

步骤如下：

```
#创建目录
[root@Superintendent2 ~]# mkdir flume
#跳转到 flume 目录
[root@Superintendent2 ~]# cd flume/
#创建 option_agent2 文件
[root@Superintendent2 flume]# vi option_agent2
#添加以下配置信息
#配置 Source 的名称
a1.sources = r1
#配置 Sink 的名称
a1.sinks = k1
#配置 Channel 的名称
a1.channels = c1
#配置 Source 的类型是 Avro
a1.sources.r1.type = avro
#配置接收 Source 数据的节点的名称
a1.sources.r1.bind = Superintendent2
#配置接收 Source 数据的节点的端口
a1.sources.r1.port = 60000
#配置 Sink 的类型为 Logger
a1.sinks.k1.type = logger
#配置缓存数据的 Channel 类型为内存
a1.channels.c1.type = memory
#配置 Channel 的容量
a1.channels.c1.capacity = 1000
#配置 Channel 每次提交事务的记录条数
a1.channels.c1.transactionCapacity = 100
#将 Source 和 Channel 连接起来
a1.sources.r1.channels = c1
#将 Sink 和 Channel 连接起来
a1.sinks.k1.channel = c1
```

④首先启动 Superintendent2 上的 Flume，命令如下：

```
#Superintendent2 上的 Flume 启动
[root@Superintendent2 flume]# flume-ng agent -n a1 -conf-file
   option_agent2 -Dflume.root.logger = INFO,console
```

启动结果如下：

```
1. 20/04/03 23:44:32 INFO node.PollingPropertiesFileConfiguration
   Provider:Configuration provider starting
2.20/04/03 23:44:32 INFO node.PollingPropertiesFileConfiguration
   Provider:Reloading configuration file:option_agent2
3.20/04/03 23:44:32 INFO conf.FlumeConfiguration:Added sinks:k1 A-
   gent:a1
4.20/04/03 23:44:32 INFO conf.FlumeConfiguration:Processing:k1
5.20/04/03 23:44:32 INFO conf.FlumeConfiguration:Processing:k1
6.20/04/03 23:44:32 INFO conf.FlumeConfiguration:Post - valida-
   tion flume configuration contains configuration for agents:[a1]
7.20/04/03 23:44:32 INFO node.AbstractConfigurationProvider:Cre-
   ating channels
8.20/04/03 23:44:32 INFO channel.DefaultChannelFactory:Creating
   instance of channel c1 type memory
9.20/04/03 23:44:32 INFO node.AbstractConfigurationProvider:Cre-
   ated channel c1
10.20/04/03 23:44:32 INFO source.DefaultSourceFactory:Creating
   instance of source r1,type avro
11.20/04/03 23:44:33 INFO sink.DefaultSinkFactory:Creating in-
   stance of sink:k1,type:logger
12.20/04/03 23:44:33 INFO node.AbstractConfigurationProvider:
   Channel c1 connected to[r1,k1]
13.20/04/03 23:44:33 INFO node.Application:Starting new configu-
   ration:{ sourceRunners:{r1 = EventDrivenSourceRunner:{ source:
   Avro source r1:{ bindAddress:Superintendent2,port:60000 } }}
   sinkRunners:{k1 = SinkRunner:{ policy:org.apache.flume.sink.
   DefaultSinkProcessor@77310e1d counterGroup:{ name:null count-
   ers:{} } }} channels:{c1 =org.apache.flume.channel. MemoryChan-
   nel{name:c1}} }
14.20/04/03 23:44:33 INFO node.Application:Starting Channel c1
15.20/04/03 23:44:33 INFO instrumentation.MonitoredCounterGroup:
   Monitored counter group for type:CHANNEL,name:c1:Successfully
   registered new MBean.
16.20/04/03 23:44:33 INFO instrumentation. MonitoredCounterGroup:
   Component type:CHANNEL,name:c1 started
17.20/04/03 23:44:33 INFO node.Application:Starting Sink k1
```

```
20/04/03 23:44:33 INFO node.Application:Starting Source r1
20/04/03 23:44:33 INFO source.AvroSource:Starting Avro source r1:{ bindAddress:Superintendent2,port:60000 }...
20/04/03 23:44:33 INFO instrumentation.MonitoredCounterGroup:Monitored counter group for type:SOURCE,name:r1:Successfully registered new MBean.
20/04/03 23:44:33 INFO instrumentation.MonitoredCounterGroup:Component type:SOURCE,name:r1 started
20/04/03 23:44:33 INFO source.AvroSource:Avro source r1 started.
```

⑤当 Superintendent2 上的 Flume 启动完成之后，再在 Superintendent1 上启动 Flume。

```
#在 Superintendent1 上启动 Flume
[root@Superintendent1 flume]# flume-ng agent -n a1 -conf-file option_agent1 -Dflume.root.logger=INFO,console
```

启动结果如下：

```
20/04/03 23:56:23 INFO node.PollingPropertiesFileConfigurationProvider:Configuration provider starting
20/04/03 23:56:23 INFO node.PollingPropertiesFileConfigurationProvider:Reloading configuration file:option_agent1
20/04/03 23:56:23 INFO conf.FlumeConfiguration:Added sinks:k1 Agent:a1
20/04/03 23:56:23 INFO conf.FlumeConfiguration:Processing:k1
20/04/03 23:56:23 INFO conf.FlumeConfiguration:Processing:k1
20/04/03 23:56:23 INFO conf.FlumeConfiguration:Processing:k1
20/04/03 23:56:23 INFO conf.FlumeConfiguration:Processing:k1
20/04/03 23:56:23 INFO conf.FlumeConfiguration:Post-validation flume configuration contains configuration for agents:[a1]
20/04/03 23:56:23 INFO node.AbstractConfigurationProvider:Creating channels
20/04/03 23:56:23 INFO channel.DefaultChannelFactory:Creating instance of channel c1 type memory
20/04/03 23:56:23 INFO node.AbstractConfigurationProvider:Created channel c1
20/04/03 23:56:23 INFO source.DefaultSourceFactory:Creating instance of source r1,type netcat
```

```
13.20/04/03 23:56:23 INFO sink.DefaultSinkFactory:Creating instance of sink:k1,type:avro
14.20/04/03 23:56:23 INFO sink.AbstractRpcSink:Connection reset is set to 0. Will not reset connection to next hop
15.20/04/03 23:56:23 INFO node.AbstractConfigurationProvider:Channel c1 connected to[r1,k1]
16.20/04/03 23:56:23 INFO node.Application:Starting new configuration:{ sourceRunners:{r1 = EventDrivenSourceRunner:{ source:org.apache.flume.source.NetcatSource{name:r1,state:IDLE} }} sinkRunners:{k1 = SinkRunner:{ policy:org.apache.flume.sink.DefaultSinkProcessor@503f76f8 counterGroup:{ name:null counters:{} } }} channels:{c1 = org.apache.flume.channel. MemoryChannel{name:c1}} }
17.20/04/03 23:56:23 INFO node.Application:Starting Channel c1
18.20/04/03 23:56:23 INFO instrumentation.MonitoredCounterGroup:Monitored counter group for type:CHANNEL,name:c1:Successfully registered new MBean.
19.20/04/03 23:56:23 INFO instrumentation. MonitoredCounterGroup:Component type:CHANNEL,name:c1 started
20.20/04/03 23:56:23 INFO node.Application:Starting Sink k1
21.20/04/03 23:56:23 INFO node.Application:Starting Source r1
22.20/04/03 23:56:23 INFO source.NetcatSource:Source starting
23.20/04/03 23:56:23 INFO sink.AbstractRpcSink:Starting RpcSink k1 { host:Superintendent2,port:60000 }...
24.20/04/03 23:56:23 INFO instrumentation.MonitoredCounterGroup:Monitored counter group for type: SINK, name: k1: Successfully registered new MBean.
25.20/04/03 23:56:23 INFO instrumentation. MonitoredCounterGroup:Component type:SINK,name:k1 started
26.20/04/03 23:56:23 INFO sink.AbstractRpcSink:Rpc sink k1:Building RpcClient with hostname:Superintendent2,port:60000
27.20/04/03 23:56:23 INFO sink.AvroSink:Attempting to create Avro Rpc client.
28.20/04/03 23:56:23 WARN api.NettyAvroRpcClient:Using default maxIOWorkers
```

```
29. 20/04/03 23:56:23 INFO source.NetcatSource:Created serverSock-
    et:sun.nio.ch.ServerSocketChannelImpl[/192.168.3.191:44444]
30.20/04/03 23:56:24 INFO sink.AbstractRpcSink:Rpc sink k1 started
```

⑥当两台 Flume 都启动完成之后，通过 YYR 的 Telnet 进行数据的输入。

```
1.#telnet 发送数据
2.[root@YYR ~]# telnet Superintendent1 44444
3.Trying 192.168.3.191...
4.Connected to Superintendent1.
5.Escape character is '^]'.
6.#发送数据到 Superintendent1
7.flume_agent
8.OK
```

⑦查看 Superintendent2 的 Flume 的控制台信息，发现刚刚 Telnet 发送的 Flume_agent 字符串已经正常显示。

```
1. 20/04/04 00:01:41 INFO ipc.NettyServer:[id:0x95370aa2,/
   192.168.3.191:39672 => /192.168.3.192:60000]OPEN
2.20/04/04 00:01:41 INFO ipc.NettyServer:[id:0x95370aa2,/
   192.168.3.191:39672 => /192.168.3.192:60000] BOUND:/192.168.
   3.192:600
3.0020/04/04 00:01:41 INFO ipc.NettyServer:[id:0x95370aa2,/
   192.168.3.191:39672 => /192.168.3.192:60000] CONNECTED:/192.
   168.3.191:39672
4.#输出数据
5.20/04/04 00:01:55 INFO sink.LoggerSink:Event:{ headers:{} body:
   66 6C 75 6D 65 5F 61 67 65 6E 74 0D flume_agent. }
```

3. 场景三

在企业的真实生产环境中，不可能所有的数据都通过 Telnet 的方式发送，而更多的是将数据存储到某一文件中，然后监控文件的增量数据，将增量的数据输入 Flume 中。流程图如图 5 – 32 所示。

操作步骤如下：

①在 Superintendent2 的/root/flume 文件夹下创建 option_exec 文件，并添加配置。

```
1.[root@Superintendent2 flume]# vi option_exec
2.#配置 Source 的名称
```

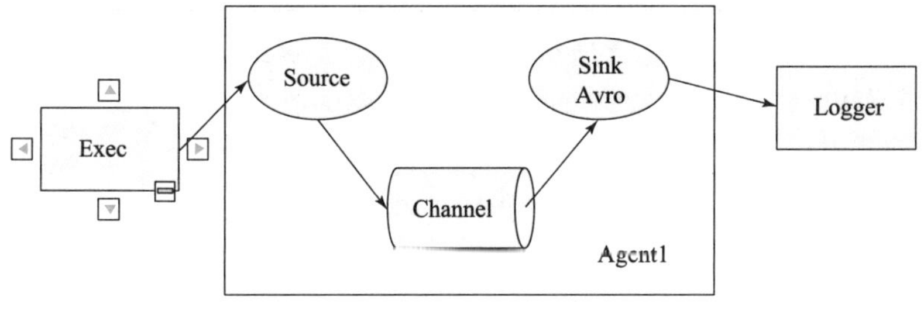

图 5-32　传输增量数据到 Flume 的执行流程

场景三:FLUME
应用案例

```
a1.sources = r1
#配置 Sink 的名称
a1.sinks = k1
#配置 Channel 的名称
a1.channels = c1
# 配置 Source 的类型为 Exec
a1.sources.r1.type = exec
#配置监控文件的目录名称
a1.sources.r1.command = tail -F /root/flume.exec.log
# 配置 Sink 的类型为 Logger
a1.sinks.k1.type = logger
# 配置 Channel 的类型为内存
a1.channels.c1.type = memory
#配置 Channel 的容量
a1.channels.c1.capacity = 1000
#配置 Channel 每次进行事务提交的容量
a1.channels.c1.transactionCapacity = 100
#将 Source 和 Channel 连接起来
a1.sources.r1.channels = c1
#将 Sink 和 Channel 连接起来
a1.sinks.k1.channel = c1
```

②在 Superintendent2 上启动 Flume。

```
[root@Superintendent2 flume]# flume-ng agent -n a1 -conf-file
option_exec -Dflume.root.logger = INFO,console
```

启动结果如下：

```
20/04/04 00:27:05 INFO node.PollingPropertiesFileConfiguration
Provider:Configuration provider starting
```

```
2. 20/04/04 00:27:05 INFO node.PollingPropertiesFileConfiguration
   Provider:Reloading configuration file:option_exec
3.20/04/04 00:27:05 INFO conf.FlumeConfiguration:Added sinks:k1
   Agent:a1
4.20/04/04 00:27:05 INFO conf.FlumeConfiguration:Processing:k1
5.20/04/04 00:27:05 INFO conf.FlumeConfiguration:Processing:k1
6.20/04/04 00:27:05 INFO conf.FlumeConfiguration:Post-valida-
   tion flume configuration contains configuration for agents:[a1]
7.20/04/04 00:27:05 INFO node.AbstractConfigurationProvider:Cre-
   ating channels
8.20/04/04 00:27:05 INFO channel.DefaultChannelFactory:Creating
   instance of channel c1 type memory
9.20/04/04 00:27:06 INFO node.AbstractConfigurationProvider:Cre-
   ated channel c1
10. 20/04/04 00:27:06 INFO source.DefaultSourceFactory:Creating
    instance of source r1,type exec
11.20/04/04 00:27:06 INFO sink.DefaultSinkFactory:Creating in-
    stance of sink:k1,type:logger
12.20/04/04 00:27:06 INFO node.AbstractConfigurationProvider:
    Channel c1 connected to[r1,k1]
13.20/04/04 00:27:06 INFO node.Application:Starting new configu-
    ration:{sourceRunners:{r1=EventDrivenSourceRunner:{source:
    org.apache.flume.source.ExecSource{name:r1,state:IDLE}}}
    sinkRunners:{k1=SinkRunner:{policy:org.apache.flume.sink.
    DefaultSinkProcessor@15a54d counterGroup:{name:null count-
    ers:{}}}} channels:{c1=org.apache.flume.channel.MemoryCh-
    annel{name:c1}}}
14.20/04/04 00:27:06 INFO node.Application:Starting Channel c1
15.20/04/04 00:27:06 INFO instrumentation.MonitoredCounterGroup:
    Monitored counter group for type:CHANNEL,name:c1:Successfully
    registered new MBean.
16.20/04/04 00:27:06 INFO instrumentation.MonitoredCounterGroup:
    Component type:CHANNEL,name:c1 started
17.20/04/04 00:27:06 INFO node.Application:Starting Sink k1
18.20/04/04 00:27:06 INFO node.Application:Starting Source r1
19.20/04/04 00:27:06 INFO source.ExecSource:Exec source starting
    with command:tail -F/root/flume.exec.log
```

```
20. 20/04/04 00:27:06 INFO instrumentation.MonitoredCounterGroup:
    Monitored counter group for type: SOURCE,name: r1: Successfully
    registered new MBean.
21.20/04/04 00:27:06 INFO instrumentation. MonitoredCounterGroup:
    Component type:SOURCE,name:r1 started
```

③向监控的文件中追加内容。

```
1.#向 flume.exec.log 中追加内容
2.[root@Superintendent2 ~]# echo "flume exec" >>flume.exec.log
3.[root@Superintendent2 ~]# echo "flume exec" >>flume.exec.log
4.[root@Superintendent2 ~]# echo "flume exec" >>flume.exec.log
5.[root@Superintendent2 ~]# echo "flume exec" >>flume.exec.log
6.[root@Superintendent2 ~]# echo "flume exec" >>flume.exec.log
```

④查看 Superintendent2 上的 Flume 的控制台输出。

```
1. 20/04/04 00:27:06 INFO instrumentation.MonitoredCounterGroup:
   Component type:SOURCE,name:r1 started
2.#Flume 会出现如下结果
3.20/04/04 00:28:32 INFO sink.LoggerSink:Event:{ headers:{} body:
   66 6C 75 6D 65 20 65 78 65 63 flume exec }
4.20/04/04 00:28:36 INFO sink.LoggerSink:Event:{ headers:{} body:
   66 6C 75 6D 65 20 65 78 65 63 flume exec }
5.20/04/04 00:28:36 INFO sink.LoggerSink:Event:{ headers:{} body:
   66 6C 75 6D 65 20 65 78 65 63 flume exec }
6.20/04/04 00:28:36 INFO sink.LoggerSink:Event:{ headers:{} body:
   66 6C 75 6D 65 20 65 78 65 63 flume exec }
7.20/04/04 00:28:36 INFO sink.LoggerSink:Event:{ headers:{} body:
   66 6C 75 6D 65 20 65 78 65 63 flume exec }
```

至止，已经从监控的文件中读取出数据，并且显示结果正确。

4. 场景四

在本场景中，将学习把结果存储到 HDFS 中。这种应用场景在企业中使用较多，在大数据的应用场景中，大部分数据都将长期存储到 HDFS 中。流程如图 5-32 所示。

从图 5-33 可以看出，通过监控某一个文件的追加内容来读取数据，并且将结果存储到 HDFS 中。

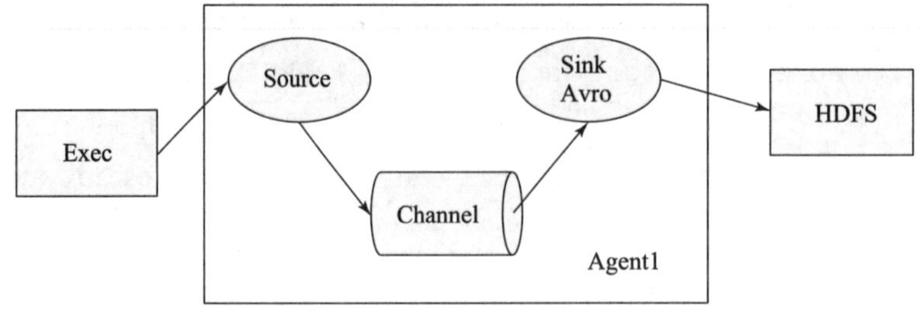

图 5-33　将数据存储到 HDFS 中的流程

操作步骤如下:

①在 Superintendent2 的/root/flume 文件夹下创建 option_exec_hdfs 文件，并添加配置。

```
#创建配置文件
[root@Superintendent2 flume]# vi option_exec_hdfs
#创建 Source 的名称
a1.sources = r1
#创建 Sink 的名称
a1.sinks = k1
#创建 Channel 的名称
a1.channels = c1
# 配置 Source 的源是 Exec
a1.sources.r1.type = exec
#配置监控的文件的名称
a1.sources.r1.command = tail -F /root/hdfs.log
# 配置 Sink 的类型
a1.sinks.k1.type = hdfs
#配置存储数据的 HDFS 的路径名称
a1.sinks.k1.hdfs.path = hdfs://YYR:9000/flume/% Y - % m - % d/%
H% M
##每隔 60 s 或者文件超过 10 MB 时产生新文件
# HDFS 有多少条消息时新建文件
a1.sinks.k1.hdfs.rollCount = 0
# HDFS 创建多长时间新建文件
a1.sinks.k1.hdfs.rollInterval = 60
# HDFS 多大时新建文件
a1.sinks.k1.hdfs.rollSize = 10240
# 当目前被打开的临时文件在该参数指定的时间(秒)内没有任何数据写入,则将
该临时文件关闭并重命名成目标文件
a1.sinks.k1.hdfs.idleTimeout = 3
```

```
#配置存储到 HDFS 的文件类型
a1.sinks.k1.hdfs.fileType = DataStream
#配置使用本地时间戳
a1.sinks.k1.hdfs.useLocalTimeStamp = true
## 每5 min 生成一个目录:
#是否启用时间上的"舍弃",这里的"舍弃"类似于"四舍五入"。如果启用,则会影响除了%t 外的所有时间表达式
a1.sinks.k1.hdfs.round = true
# 时间上进行"舍弃"的值
a1.sinks.k1.hdfs.roundValue = 5
# 时间上进行"舍弃"的单位,包含 second,minute,hour
a1.sinks.k1.hdfs.roundUnit = minute
# 配置 Channel 的类型为内存
a1.channels.c1.type = memory
#配置 Channel 的容量
a1.channels.c1.capacity = 1000
#配置每次提交事务的容量
a1.channels.c1.transactionCapacity = 100
#将 Source 和 Channel 连接起来
a1.sources.r1.channels = c1
#将 Sink 和 Channel 连接起来
a1.sinks.k1.channel = c1
```

②在 Superintendent2 上启动 Flume。

```
[root@Superintendent2 flume]# flume-ng agent -n a1 -conf-file option_exec_file -Dflume.root.logger = INFO,console
```

③向监控的文件中追加内容。

```
20/04/04 00:59:19 INFO node.PollingPropertiesFileConfigurationProvider:Configuration provider starting
20/04/04 00:59:19 INFO node.PollingPropertiesFileConfigurationProvider:Reloading configuration file:option_exec_hdfs
20/04/04 00:59:19 INFO conf.FlumeConfiguration:Processing:k1
20/04/04 00:59:19 INFO conf.FlumeConfiguration:Added sinks:k1 Agent:a1
20/04/04 00:59:19 INFO conf.FlumeConfiguration:Processing:k1
20/04/04 00:59:19 INFO conf.FlumeConfiguration:Processing:k1
20/04/04 00:59:19 INFO conf.FlumeConfiguration:Processing:k1
```

```
8. 20/04/04 00:59:19 INFO conf.FlumeConfiguration:Processing:k1
9.20/04/04 00:59:19 INFO conf.FlumeConfiguration:Processing:k1
10. 20/04/04 00:59:19 INFO conf.FlumeConfiguration:Processing:k1
11.20/04/04 00:59:19 INFO conf.FlumeConfiguration:Processing:k1
12.20/04/04 00:59:19 INFO conf.FlumeConfiguration:Processing:k1
13.20/04/04 00:59:19 INFO conf.FlumeConfiguration:Processing:k1
14.20/04/04 00:59:19 INFO conf.FlumeConfiguration:Processing:k1
15.20/04/04 00:59:19 INFO conf.FlumeConfiguration:Processing:k1
16.20/04/04 00:59:19 INFO conf.FlumeConfiguration: Post - vali-
   dation flume configuration contains configuration for agents:[a1]
17.20/04/04 00:59:19 INFO node.AbstractConfigurationProvider:
   Creating channels
18.20/04/04 00:59:19 INFO channel.DefaultChannelFactory:Creating
   instance of channel c1 type memory
19.20/04/04 00:59:19 INFO node.AbstractConfigurationProvider:
   Created channel c1
20.20/04/04 00:59:19 INFO source.DefaultSourceFactory:Creating
   instance of source r1,type exec
21.20/04/04 00:59:19 INFO sink.DefaultSinkFactory:Creating in-
   stance of sink:k1,type:hdfs
22.20/04/04 00:59:19 INFO node.AbstractConfigurationProvider:
   Channel c1 connected to[r1,k1]
23.20/04/04 00:59:19 INFO node.Application:Starting new configu-
   ration:{ sourceRunners:{r1 = EventDrivenSourceRunner:{ source:
   org.apache.flume.source.ExecSource{name: r1, state: IDLE} }}
   sinkRunners:{k1 = SinkRunner:{ policy:org.apache.flume.sink.
   DefaultSinkProcessor@71d09cc1 counterGroup:{ name:null count-
   ers:{} } }} channels:{c1 = org.apache.flume.channel. MemoryCh-
   annel{name:c1}} }
24.20/04/04 00:59:19 INFO node.Application:Starting Channel c1
25.20/04/04 00:59:19 INFO instrumentation.MonitoredCounterGroup:
   Monitored counter group for type:CHANNEL,name:c1:Successfully
   registered new MBean.
26. 20/04/04 00:59:19 INFO instrumentation. MonitoredCounterG-
   roup:Component type:CHANNEL,name:c1 started
27.20/04/04 00:59:19 INFO node.Application:Starting Sink k1
28. 20/04/04 00:59:19 INFO node.Application:Starting Source r1
```

```
29. 20/04/04 00:59:19 INFO source.ExecSource:Exec source starting with command:tail -F/root/hdfs.log
30.20/04/04 00:59:19 INFO instrumentation.MonitoredCounterGroup: Monitored counter group for type: SINK, name: k1: Successfully registered new MBean.
31.20/04/04 00:59:19 INFO instrumentation. MonitoredCounterGroup: Component type:SINK,name:k1 started
32.20/04/04 00:59:19 INFO instrumentation.MonitoredCounterGroup: Monitored counter group for type: SOURCE, name: r1: Successfully registered new MBean.
33.20/04/04 00:59:19 INFO instrumentation. MonitoredCounterGroup: Component type:SOURCE,name:r1 started
```

④查看 Superintendent2 上的 Flume 的控制台输出。

```
1. 20/04/04 01:07:11 INFO hdfs.HDFSDataStream: Serializer = TEXT, UseRawLocalFileSystem = false
2.20/04/04 01:07:11 INFO hdfs.BucketWriter: Creating hdfs://YYR:9000/flume/2020-04-04/0105/FlumeData.1554311231460.tmp
3.20/04/04 01:07:20 INFO hdfs.BucketWriter: Closing idle bucket-Writer hdfs://YYR:9000/flume/2020-04-04/0105/FlumeData.1554311231460.tmp at 1554311240176
4.20/04/04 01:07:20 INFO hdfs.BucketWriter: Closing hdfs://YYR:9000/flume/2020-04-04/0105/FlumeData.1554311231460.tmp
5.20/04/04 01:07:20 INFO hdfs.BucketWriter: Renaming hdfs://YYR:9000/flume/2020-04-04/0105/FlumeData.1554311231460.tmp to hdfs://master:9000/flume/2020-04-04/0105/FlumeData.1554311231460
6.20/04/04 01:07:20 INFO hdfs.HDFSEventSink: Writer callback called.
```

由图 5－34 可见，HDFS 中已经存在文件，表示文件上传成功。

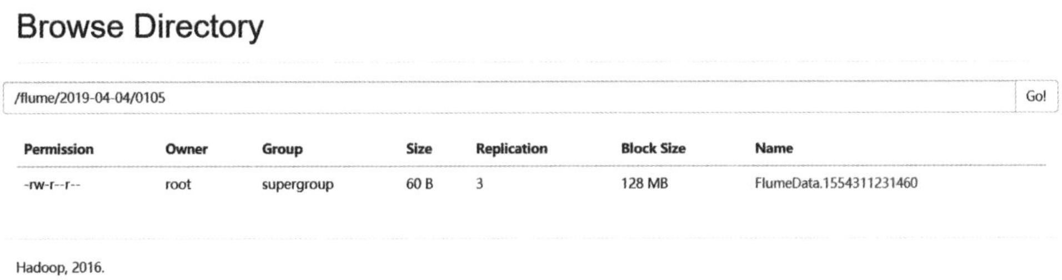

Browse Directory

/flume/2019-04-04/0105　Go!

Permission	Owner	Group	Size	Replication	Block Size	Name
-rw-r--r--	root	supergroup	60 B	3	128 MB	FlumeData.1554311231460

Hadoop, 2016.

图 5－34　查看 HDFS 目录

学习笔记

项目六
Sqoop——海量数据传输工具使用

项目描述

在本项目中，数据分析组会通过 Hive 来对相应指标进行分析，分析之后的结果会暂存在 Hive 的表中，但数据最终要以可视化的方式呈现，所以需要将 Hive 中存储结果的表的数据导出到 MySQL 中，从而方便数据可视化组从 MySQL 中获取数据，进行数据的可视化（图表展示）操作。

项目分析

主要通过以下几个任务来完成海量数据传输工具 Sqoop 的学习：

任务 1：Sqoop 组件的安装。

任务 2：Sqoop 的数据导入与导出。

在数据的导出阶段，采用 Apache Sqoop 开源工具进行数据的导出，如图 6－1 所示。

Sqoop 的产生主要源于大部分使用 Hadoop 技术处理大数据业务的企业，它们有大量的数据存储在传统的关系型数据库（RDBMS）中；由于缺乏工具的支持，对 Hadoop 和传统数据库系统中的数据进行传输是一件十分困难的事情。基于这两个因素，急需一个在 RDBMS 与 Hadoop 之间进行数据传输的工具。

Sqoop 是一个用于在 Hadoop 和关系型数据库之间传输数据的工具。可以使用 Sqoop 将数据从 RDBMS（比如 MySQL 或者 Oracle）导出到 Hadoop 分布式文件系统（HDFS）中，然后数据在 Hadoop MapReduce 上转换，并被导出到 RDBMS 中。

Sqoop 自动实现了上面提及的很多过程。Sqoop 使用 MapReduce 来导入和导出数据，这样既能提供并行化操作，又能提高容错能力。

Sqoop 是连接传统关系型数据库和 Hadoop 的纽带，它包括两方面内容：

①将 RDBMS 的数据导出到 Hadoop 及其相关的系统中，如 Hive 和 HBase。

②将数据从 Hadoop 系统里抽取并导出到 RDBMS。

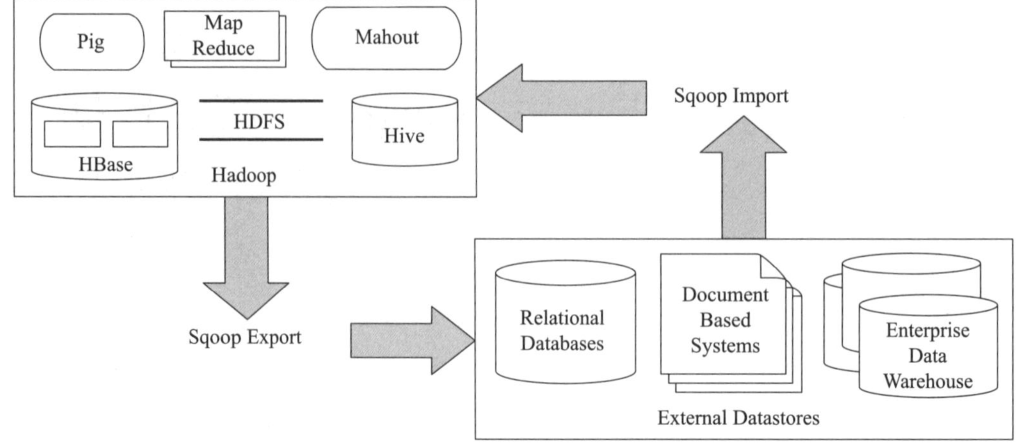

图6－1　Sqoop的工作流程

任务1　Sqoop组件的安装 

安装SQOOP

1. 安装Sqoop

①从官方网站下载Sqoop的安装包，如图6－2所示。单击图中框住的部分，选择对应的Sqoop版本进行下载。

②下载完成之后，上传Sqoop的压缩文件到Superintendent2中。

③解压Sqoop的压缩文件到/usr/local/目录下。

```
[root@Superintendent2 ~]# tar -zxvf /tmp/sqoop-1.4.7.bin__hadoop-2.6.0.tar.gz -C/usr/local/
sqoop-1.4.7.bin__hadoop-2.6.0/
sqoop-1.4.7.bin__hadoop-2.6.0/CHANGELOG.txt
... （此处省略解压过程）
```

④将文件夹的名称Sqoop改为sqoop。

```
[root@Superintendent2 ~]# ll/usr/local/
总用量 0
drwxr-xr-x. 2 root root   6 11月  5 2016 bin
drwxr-xr-x. 2 root root   6 11月  5 2016 etc
drwxr-xr-x  7 root root 162 5月  15 09:40 flume
drwxr-xr-x. 2 root root   6 11月  5 2016 games
```

Sqoop ▾　Project Information ▾　Releases ▾　Documentation ▾　ASF ▾　External Links ▾

Apache / Sqoop /　　Last Published: 2019-01-18

Apache Sqoop

Apache Sqoop(TM) is a tool designed for efficiently transferring bulk data between Apache Hadoop and structured datastores such as relational databases.

Sqoop successfully graduated from the Incubator in March of 2012 and is now a Top-Level Apache project: More information

Latest stable release is 1.4.7 (download, documentation). Latest cut of Sqoop2 is 1.99.7 (download, documentation). Note that 1.99.7 is not compatible with 1.4.7 and not feature complete, it is not intended for production deployment.

Download

Download a release of Sqoop from a nearby mirror.

Sqoop source code is available on both Gitbox and GitHub.

You might clone the repository using one of the following commands:

git clone https://gitbox.apache.org/repos/asf/sqoop.git
git clone https://github.com/apache/sqoop.git

Use one of the following links to browse the repository online:

https://gitbox.apache.org/repos/asf?p=sqoop.git
https://github.com/apache/sqoop

图 6-2　官网下载 Sqoop

```
drwxr-xr-x  12 root root 184 4月  28 10:02 hadoop
drwxr-xr-x   9 root root 170 5月   9 13:30 hive
drwxr-xr-x.  2 root root   6 11月  5 2016 include
drwxr-xr-x.  2 root root   6 11月  5 2016 lib
drwxr-xr-x.  2 root root   6 11月  5 2016 lib64
drwxr-xr-x.  2 root root   6 11月  5 2016 libexec
drwxr-xr-x.  2 root root   6 11月  5 2016 sbin
drwxr-xr-x.  5 root root  49 1月  10 10:03 share
drwxr-xr-x   9 1000 1000 318 12月 19 2017 sqoop-1.4.7.bin__hadoop-2.6.0
drwxr-xr-x.  2 root root   6 11月  5 2016 src
# 修改 Sqoop 文件夹名称
[root@Superintendent2 ~]# mv /usr/local/sqoop-1.4.7.bin__hadoop-2.6.0 //usr/local/sqoop
```

```
[root@Superintendent2 ~]# ll/usr/local/
总用量 0
drwxr-xr-x.  2 root root   6 11 月  5 2016 bin
drwxr-xr-x.  2 root root   6 11 月  5 2016 etc
drwxr-xr-x   7 root root 162 5 月  15 09:40 flume
drwxr-xr-x.  2 root root   6 11 月  5 2016 games
drwxr-xr-x  12 root root 184 4 月  28 10:02 hadoop
drwxr-xr-x   9 root root 170 5 月   9 13:30 hive
drwxr-xr-x.  2 root root   6 11 月  5 2016 include
drwxr-xr-x.  2 root root   6 11 月  5 2016 lib
drwxr-xr-x.  2 root root   6 11 月  5 2016 lib64
drwxr-xr-x.  2 root root   6 11 月  5 2016 libexec
drwxr-xr-x.  2 root root   6 11 月  5 2016 sbin
drwxr-xr-x.  5 root root  49 1 月  10 10:03 share
drwxr-xr-x   9 1000 1000 318 12 月 19 2017 sqoop
drwxr-xr-x.  2 root root   6 11 月  5 2016 src
```

⑤配置 Sqoop 的环境变量。

```
[root@Superintendent2 ~]# vi/etc/profile

# 文件结尾处增加如下内容
# set sqoop enviroment
export SQOOP_HOME = /usr/local/sqoop
export PATH = $PATH: ${SQOOP_HOME} /bin
```

⑥执行命令，使刚刚修改的环境变量生效。

```
[root@Superintendent2 ~]# source/etc/profile
```

2. 验证 Sqoop

通常情况下，上述步骤完成后，Sqoop 便安装成功。现在来验证 Sqoop 是否安装成功，执行以下命令即可。

```
[root@Superintendent2 ~]# sqoop version
Warning:/usr/local/sqoop/../hbase does not exist! HBase imports will fail.
Please set $HBASE_HOME to the root of your HBase installation.
Warning:/usr/local/sqoop/../hcatalog does not exist! HCatalog jobs will fail.
```

```
Please set $HCAT_HOME to the root of your HCatalog installation.
Warning:/usr/local/sqoop/../accumulo does not exist! Accumulo imports will fail.
Please set $ACCUMULO_HOME to the root of your Accumulo installation.
Warning:/usr/local/sqoop/../zookeeper does not exist! Accumulo imports will fail.
Please set $ZOOKEEPER_HOME to the root of your Zookeeper installation.
20/04/04 05:14:26 INFO sqoop.Sqoop:Running Sqoop version:1.4.6
Sqoop 1.4.6
git commit id c0c5a81723759fa575844a0a1eae8f510fa32c25
Compiled by root on Mon Apr 27 14:38:36 CST 2015
```

由上述代码可见，已经正确打印了 Sqoop 的版本信息，说明 Sqoop 安装成功。但是在版本的前面出现了一些警告信息，现在将这些警告信息消除。

```
#切换到 Sqoop 的 bin 目录
[root@Superintendent2 ~]# cd/usr/local/sqoop/bin/
#展示当前目录的文件
[root@Superintendent2 bin]# ls
configure-sqoop sqoop-codegen sqoop-help sqoop-job sqoop-metastore
configure-sqoop.cmd sqoop-create-hive-table sqoop-import sqoop-list-databases sqoop-version
sqoop sqoop-eval sqoop-import-all-tables sqoop-list-tables start-metastore.sh
sqoop.cmd sqoop-export sqoop-import-mainframe sqoop-merge stop-metastore.sh
#修改 configure-sqoop 文件
[root@Superintendent2 bin]# vi configure-sqoop
#从第128行开始,看到如下信息
if[!-d "${HBASE_HOME}"];then
 echo "Warning: $HBASE_HOME does not exist! HBase imports will fail."
echo 'Please set $HBASE_HOME to the root of your HBase installation.'
fi

## Moved to be a runtime check in sqoop.
```

```
if[!-d "${HCAT_HOME}"];then
echo "Warning: $HCAT_HOME does not exist! HCatalog jobs will fail."
echo 'Please set $HCAT_HOME to the root of your HCatalog installation.'
fi

if[!-d "${ACCUMULO_HOME}"];then
echo "Warning: $ACCUMULO_HOME does not exist! Accumulo imports will fail."
echo 'Please set $ACCUMULO_HOME to the root of your Accumulo installation.'
fi
if[!-d "${ZOOKEEPER_HOME}"];then
echo "Warning: $ZOOKEEPER_HOME does not exist! Accumulo imports will fail."
echo 'Please set $ZOOKEEPER_HOME to the root of your Zookeeper installation.'
fi
#将这几行的内容注释起来,在每一行的前面添加#,如下所示:
## Moved to be a runtime check in sqoop.
#if[!-d "${HBASE_HOME}"];then
# echo "Warning: $HBASE_HOME does not exist! HBase imports will fail."
# echo 'Please set $HBASE_HOME to the root of your HBase installation.'
#fi

## Moved to be a runtime check in sqoop.
#if[!-d "${HCAT_HOME}"];then
# echo "Warning: $HCAT_HOME does not exist! HCatalog jobs will fail."
# echo 'Please set $HCAT_HOME to the root of your HCatalog installation.'
#fi

#if[!-d "${ACCUMULO_HOME}"];then
# echo "Warning: $ACCUMULO_HOME does not exist! Accumulo imports will fail."
```

```
# echo 'Please set $ACCUMULO_HOME to the root of your Accumulo installation.'
#fi
#if[!-d "${ZOOKEEPER_HOME}"];then
# echo "Warning: $ZOOKEEPER_HOME does not exist! Accumulo imports will fail."
# echo 'Please set $ZOOKEEPER_HOME to the root of your Zookeeper installation.'
#fi
#保存退出后,再次运行 sqoop version,看到如下信息:
[root@Superintendent2 bin]# sqoop version
20/04/04 05:22:42 INFO sqoop.Sqoop:Running Sqoop version:1.4.7
Sqoop 1.4.7
git commit id c0c5a81723759fa575844a0a1eae8f510fa32c25
Compiled by root on Mon Apr 27 14:38:36 CST 2015
#如上所示,警告信息已经消除
```

3. JDBC 配置

配置 JDBC

Sqoop 的主要命令如下:

```
[root@Superintendent2 bin]# sqoop help
20/04/04 05:31:13 INFO sqoop.Sqoop:Running Sqoop version:1.4.7
usage:sqoop COMMAND[ARGS]

Available commands:
#获取数据库中某张表的数据
codegen Generate code to interact with database records
#创建 Hive 表
create-hive-table Import a table definition into Hive
#查看 SQL 的执行结果
eval Evaluate a SQL statement and display the results
#导出 HDFS 数据到数据库表
export Export an HDFS directory to a database table
#帮助命令
help List available commands
#导出数据到 HDFS
import Import a table from a database to HDFS
#导出某个数据库的所有表到 HDFS
```

```
import - all - tables Import tables from a database to HDFS
#导出主机数据集到 HDFS
import - mainframe Import datasets from a mainframe server to HDFS
#将任务保存为 job
job Work with saved jobs
#展示某个数据库的所有数据库名称
list - databases List available databases on a server
#展示某个数据库的所有表
list - tables List available tables in a database
#将 HDFS 上不同目录的数据合并在一起,并存放在指定目录
merge Merge results of incremental imports
#记录 Sqoop job 的元数据信息
metastore Run a standalone Sqoop metastore
#查看 Sqoop 的版本信息
version Display version information

See 'sqoop help COMMAND' for information on a specific command.
```

list - databases 和 list - tables 命令用来操作数据库，首先需要将 MySQL 的驱动包拷贝到 lib 目录下。

```
#将 MySQL 的驱动包上传到/root/目录下
#展示当前目录下的内容
[root@Superintendent2 ~]# ls
mysql - connector - java - 8.0.16 - bin.jar
#拷贝 MySQL 驱动包到 Sqoop 的 lib 目录下
[root@Superintendent2 ~]# cp mysql - connector - java - 8.0.16 - bin.jar /usr/local/sqoop/lib/
```

通过 Sqoop 的官网可以找到连接 MySQL 的对应参数，见表 6 - 1。

表 6 - 1　连接 MySQL 的对应参数

参数名称	参数描述
-- connect < jdbc - uri >	指定 JDBC 的 URL
-- connection - manager < class - name >	指定 JDBC 的连接管理类
-- driver < class - name >	指定驱动类的名称
-- hadoop - mapred - home < dir >	指定 MapReduce 的目录
-- help	帮助手册

续表

参数名称	参数描述
-- password - file	从文件中读取密码
- P	从控制台读取密码
password < password >	直接在命令后添加密码
-- username	连接 MySQL 的用户名
-- verbose	执行 Sqoop 时打印更多的信息
-- connection - param - file	指定连接参数的配置文件
-- relaxed - isolation	设置独立的连接事务

测试 SQOOP 远程

4. 测试 Sqoop 远程连接数据库

测试能否成功连接到 MySQL 数据库中，命令如下：

```
[root@Superintendent2 ~]# sqoop list - databases -- connect jdbc:mysql://YYR:3306 --username root --password root00
20/06/10 14:29:42 INFO sqoop.Sqoop:Running Sqoop version:1.4.7
20/06/10 14:29:42 WARN tool.BaseSqoopTool:Setting your password on the command - line is insecure. Consider using - P instead.
20/06/10 14:29:42 INFO manager.MySQLManager:Preparing to use a MySQL streaming resultset.
Loading class 'com.mysql.jdbc.Driver'. This is deprecated. The new driver class is 'com.mysql.cj.jdbc.Driver'. The driver is automatically registered via the SPI and manual loading of the driver class is generally unnecessary.
information_schema
hive
mysql
performance_schema
sys
[root@Superintendent2 ~]#
```

运行程序，发现 Sqoop 能够打印 MySQL 中的数据库了，表明 Sqoop 成功连接 MySQL。

任务 2 Sqoop 的数据导入与导出 

了解 Sqoop 常用的特性，以便于后续实际的操作。

1. Sqoop 架构

Sqoop 主要由三个部分组成：Sqoop Client、HDFS/HBase/Hive、Database。通过 MapReduce 任务来传输数据，从而提供并发特性和容错。底层的数据传输是通过 MapReduce/YARN 实现的。可以参考 http://sqoop. apache. org/docs/1. 4. 6/SqoopUserGuide. html。

如图 6-3 所示，用户向 Sqoop 发起一个命令之后，这个命令会转换为一个基于 Map Task 的 MapReduce 作业。Map Task 会访问数据库的元数据信息，通过并行的 Map Task 将数据库的数据读取出来，然后导入 Hadoop 中。当然，也可以将 Hadoop 中的数据导入传统的关系型数据库中。其核心思想就是通过基于 Map Task（只有 Map）的 MapReduce 作业，实现数据的并发拷贝和传输，这样可以大大提高效率。

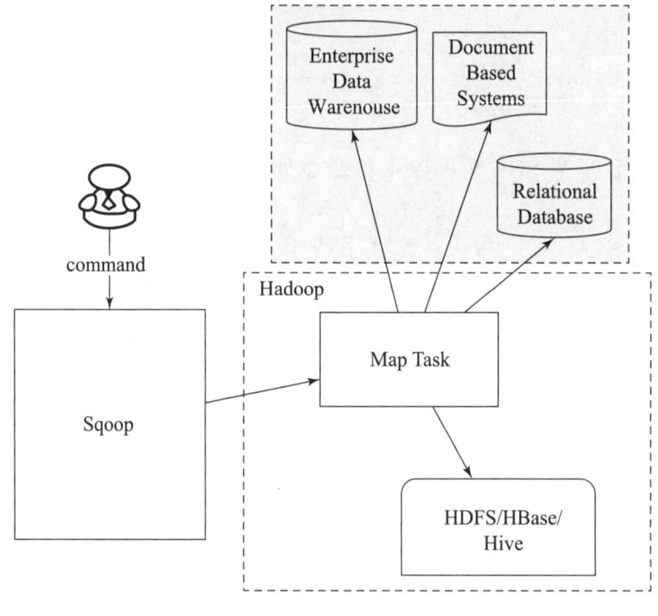

图 6-3　MapReduce 作业流程

2. Sqoop 数据导入

Sqoop 是通过 MapReduce 作业导入的，在作业中，会从表中读取一行行的记录，然后将其写入 HDFS 中。

开始导入之前，Sqoop 会通过 JDBC 获得所需的数据库元数据，例如，表的列名、数据类型等（第一步）；接着这些数据库的数据类型（varchar、number 等）会被映射成 Java 类型（String、int 等），根据这些信息，Sqoop 会生成一个与表名相同的类，用来完成反序列化的工作，保存表中的每一行记录（第二步）；Sqoop 启动 MapReduce 作业（第三步），MapReduce 中主要是对 InputFormat 和 OutputFormat 进行定制；启动的作业在 input 的过程中，会通过 JDBC 读取数据库表中的内容（第四步），这时，会使用 Sqoop 生成的类进行反序列化；最后再将这些记录写到 HDFS 中，在写入 HDFS 的过程中，同样会使用 Sqoop 生成的类进行序列化。

Sqoop 的导入作业通常不只是由一个 Map 任务完成，也就是说，每一个任务会获取表的一部分数据。

导入流程如图 6 - 4 所示。

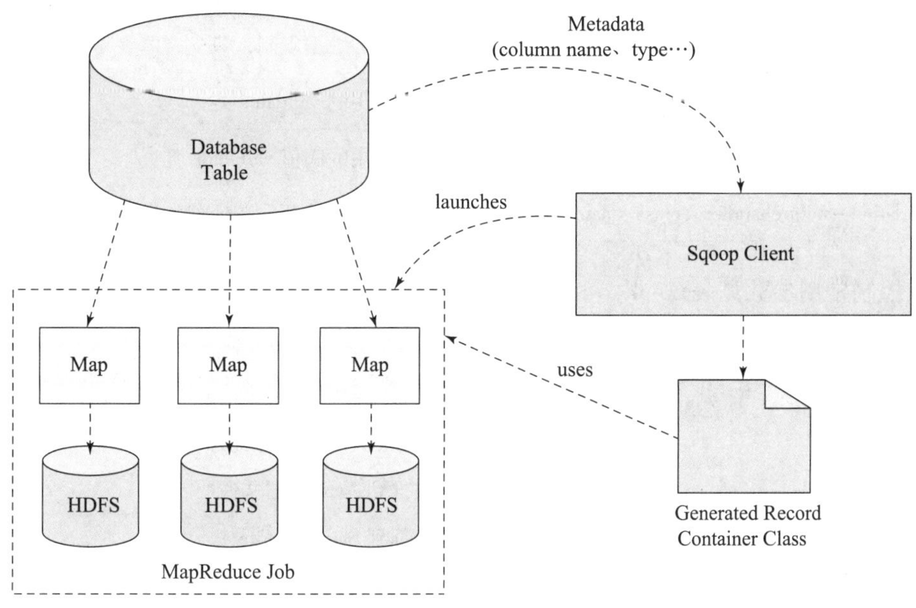

图 6 - 4　Sqoop 的导入作业操作

Sqoop 导入数据的命令如下：

```
1.sqoop import[GENERIC-ARGS][TOOL-ARGS]
```

表 6 - 2 列出了 Sqoop 导入的常用参数。

表 6 - 2　Sqoop 导入的常用参数

参数	描述
-- connect <jdbc - uri>	JDBC 连接地址
-- connection - manager <class - name>	连接管理者
-- driver <class - name>	驱动类
-- help	帮助信息
-- password - file	为包含身份验证密码的文件设置路径
- P	从命令行输入密码
-- password <password>	密码
-- username <username>	用户名
-- verbose	打印流程信息
-- connection - param - file <filename>	可选参数
-- relaxed - isolation	将连接事务隔离设置为读取未提交的映射程序

验证参数见表6－3。

表6－3　验证参数

参数	描述
－－validator	启用验证的数据复制，只支持单表复制
－－validator <class－name>	指定要使用的 validator 类
－－validation－threshold <class－name>	指定要使用的阈值 validator 类
－－validation－failurehandler <class－name>	指定要使用的验证失败处理的类

导入控制常用参数见表6－4。

表6－4　导入控制常用参数

参数	描述
－－append	追加数据到 HDFS 中已经存在的数据集
－－as－avrodatafile	导入数据到 Avro 数据文件
－－as－sequencefile	导入数据到 SequenceFile
－－as－textfile	导入格式作为无格式文档（默认）
－－boundary－query <statement>	创建被分割的边界查询语句对参数的描述
－－columns <col,col,col…>	指定导入的列
－－direct	使用导入中的快速通道，direct 模式
－－fetch－size <n>	一次性从数据库读取 n 个实例，即 n 条数据
－－inline－lob－limit <n>	设置一个内联 LOB 的最大尺寸
－m，－－num－mappers <n>	使用 n 个 Map 任务以并行方式导入
－e，－－query <statement>	可以导入一个查询的结果集，这里指定一个查询语句
－－split－by <column－name>	分割任务时的依据列
－m，－－num－mappers－－table <table－name>	使用 n 个 Map 任务以并行方式导入
－－target－dir <dir>	导入文件的存放目录
－－warehouse－dir <dir>	可以指定父级目录
－z，－－compress	启用压缩
－－compression－codec <c>	指定 Hadoop 编解码器（默认 gzip）
－－null－string <null－string>	指定导入一个空值的替换值
－－null－non－string <null－string>	非字符串类型为空时的默认值
－－delete－target－dir	如果导入目录存在，则删除

－－null－string 和－－null－non－string 参数是可选的，如果不指定，就会使用“null”。

（1）选择要导入的数据

使用 - table 参数可以指定导入的表或者视图，默认情况下，该表或视图的所有字段都会被按顺序导入进来。但是如果只需要导入该表的部分字段，或者调整导入字段的顺序，可以使用 - columns 参数。例如 -- columns "name,employee_id,jobtitle"。

也可以使用 - where 参数将符合筛选条件的记录导入进来，例如 -- where "id > 400"，则源表中只有 ID 大于 400 的记录才会被导入进来。

默认情况下，Sqoop 会根据指定的 - split - by 字段的 min、max 值对记录进行分片并导入。但是在有些情况下，如果不需要导入从 min(split - by) 到 max(split - by) 的记录，则可以使用 -- boundary - query < statement > 参数传入一个返回两个数字类型的 SQL 语句来指定导入范围。如果传入非数字类型的字段，比如 varchar 型，会报 Invalid value for getLong() 之类的错误信息。

（2）自由形态的查询导入

除了像上面那样指定 - table、- columns、- where 等参数的导入方法之外，Sqoop 还支持传入一个 SQL 语句，并把该 SQL 语句的结果导入 HDFS 中。这里需要使用 - query 或者 - e 参数。但是使用 - query 参数时，同时要指定 - target - dir 参数，如果需要并行地导入数据，即多个 Map 并行地用相同的 query 语句查询不同的数据，并最终汇总到一起，那么需要在 query 语句中增加一个 $CONDITIONS 字符串，并且指定 - split - by 字段。比如：

```
sqoop import \
 -- query 'SELECT a. * ,b. * FROM a JOIN b on (a.id == b.id) WHERE
   $CONDITIONS' \
-- split - by a.id -- target - dir/user/foo/joinresults

sqoop import \
 -- query 'SELECT a. * ,b. * FROM a JOIN b on (a.id == b.id) WHERE
   $CONDITIONS' \
- m 1 -- target - dir/user/foo/joinresults
```

第二个语句是将 Map 的个数设置成 1。这里需要注意的是，上面两个示例中使用的是单引号，如果要使用双引号，那么应该写成“$CONDITIONS”，即在前面加一个反斜杠。

在当前 Sqoop 版本中，也不能传入特别复杂的 SQL 语句，比如在 where 条件中不能有 or 操作，并且对有子查询的语句，可能会出现一些未知的错误。

（3）并行性控制

Sqoop 支持并发地从多种数据源中导入数据。可以使用 - m 或者 - num - mappers 参数指定 Map 的个数，每一个 Map 处理数据源中的一部分数据。默认情况下，Sqoop 的并发数为 4，对某些数据库来说，将并发数调整到 8 或 16 会有显著的性能提升。

但是并发调整得太高也不好，如果 Map 个数太多，超出整个集群的最大并行数，则其他的 Map 任务就需要等前面的 Map 任务执行完毕才能继续执行，这样反而会增加导入总时间。同时，如果并发数太高，会增加数据库的访问压力，所以，应该根据实际情况调整至最优的并发数。

当并发地导入数据时，Sqoop 需要有一个明确的划分数据集的指标，告诉 Sqoop 如何为每个 Map 分配处理的数据，使得既无遗漏，也无重复。默认情况下，如果指定表有主键，Sqoop 会根据主键的 max、min 值除以 Map 个数来划分数据集。比如一张表有 1 000 条记录，主键为 id，并发数为 4，那么每个 Map 上执行的 SQL 语句是 SELECT * FROM sometable WHERE id > = lo AND id < hi。对每个 Map 来说，（lo,hi）分别是（0,250），（250,500），（500,750）及（750,1001）。

但是这样会有一个数据倾斜的问题，比如 1 000 条记录大部分分布在 0 ~ 250 区间内，那么会有一个 Map 处理大量数据，而其他 Map 处理的数据特别少，这时就需要使用 -- split - by 参数手动指定一个分割字段。目前 Sqoop 不支持联合主键，那么如果导入表没有主键，或者有联合主键时，就需要手动指定分割字段。

（4）控制导入过程

默认情况下，Import 过程会使用 JDBC 提供的稳定服务去访问和连接数据库，但是有些数据库提供了一些其他的数据库访问工具来为 import 过程提供更高效的服务。比如，MySQL 提供了一个 mysqldump 工具，可以得从 MySQL 导出数据到其他系统更加快速。使用 -- direct 参数就可以指定 Sqoop 使用 mysqldump 从数据库导入数据。这个方法比 JDBC 提供的服务更加高效。

默认情况下，Sqoop 会将一个表名为 foo 的表导入到 HDFS 上一个名称为 foo 的路径下。比如，如果使用用户 someuser 来执行 Sqoop import，那么会将 foo 表导入到 HDFS 上的/user/someuser/foo/(files) 路径下。当然，也可以使用 -- warehouse - dir 参数来指定导入文件的父路径。

下面进行实际操作。

- 从 RDBMS 导入到 HDFS 中

实例 1：普通导入。

实例 1：普通导入

```
# 数据导入命令
[root@Superintendent2 ~]# sqoop import --connect jdbc:mysql://YYR:3306/mysql --username root --password root00 --table help_keyword --target-dir /user/YYR/my_help_keyword
# 数据导入命令执行过程如下
20/06/11 15:12:21 INFO sqoop.Sqoop:Running Sqoop version:1.4.7
20/06/11 15:12:21 WARN tool.BaseSqoopTool:Setting your password on the command-line is insecure. Consider using -P instead.
20/06/11 15:12:21 INFO manager.MySQLManager:Preparing to use a MySQL streaming resultset.
20/06/11 15:12:21 INFO tool.CodeGenTool:Beginning code generation
```

```
8. Loading class 'com.mysql.jdbc.Driver'. This is deprecated. The
   new driver class is 'com.mysql.cj.jdbc.Driver'. The driver is au-
   tomatically registered via the SPI and manual loading of the
   driver class is generally unnecessary.
9.20/06/11 15:12:22 INFO manager.SqlManager:Executing SQL state-
   ment:SELECT t.* FROM 'help_keyword' AS t LIMIT 1
10. 20/06/11 15:12:22 INFO manager.SqlManager:Executing SQL state-
    ment:SELECT t.* FROM 'help_keyword' AS t LIMIT 1
11.20/06/11 15:12:22 INFO orm.CompilationManager:HADOOP_MAPRED_
    HOME is/usr/local/hadoop
12. 注:/tmp/sqoop-root/compile/2b3e42a502a0c7b61a748ef35f49e00f/
    help_keyword.java 使用或覆盖了已过时的 API。
13. 注:有关的详细信息,请使用 -Xlint:deprecation 重新编译。
14. 20/06/11 15:12:26 INFO orm.CompilationManager: Writing jar
    file:/tmp/sqoop-root/compile/2b3e42a502a0c7b61a748ef35f49
    e00f/help_keyword.jar
15.20/06/11 15:12:26 WARN manager.MySQLManager:It looks like you
    are importing from mysql.
16.20/06/11 15:12:26 WARN manager.MySQLManager:This transfer can
    be faster! Use the --direct
17.20/06/11 15:12:26 WARN manager.MySQLManager:option to exercise
    a MySQL-specific fast path.
18.20/06/11 15:12:26 INFO manager.MySQLManager:Setting zero DATE-
    TIME behavior to convertToNull (mysql)
19.20/06/11 15:12:26 INFO mapreduce.ImportJobBase:Beginning im-
    port of help_keyword
20.20/06/11 15:12:26 INFO Configuration.deprecation:mapred.jar is
    deprecated. Instead,use mapreduce.job.jar
21.20/06/11 15:12:26 INFO Configuration.deprecation:mapred.map.
    tasks is deprecated. Instead,use mapreduce.job.maps
22.20/06/11 15:12:26 INFO client.RMProxy:Connecting to ResourceM-
    anager at YYR/192.168.3.190:8032
23.20/06/11 15:12:29 INFO db.DBInputFormat:Using read commited
    transaction isolation
24.20/06/11 15:12:29 INFO db.DataDrivenDBInputFormat: Bound-
    ingValsQuery:SELECT MIN('help_keyword_id'),MAX('help_keyword_
    id') FROM 'help_keyword'
```

```
20/06/11 15:12:29 INFO db.IntegerSplitter:Split size:174;Num splits:4 from:0 to:698
20/06/11 15:12:29 INFO mapreduce.JobSubmitter: number of splits:4
20/06/11 15:12:29 INFO mapreduce.JobSubmitter:Submitting tokens for job:job_1557902299589_0011
20/06/11 15:12:29 INFO impl.YarnClientImpl:Submitted application application_1557902299589_0011
20/06/11 15:12:29 INFO mapreduce.Job:The url to track the job: http://YYR:8088/proxy/application_1557902299589_0011/
20/06/11 15:12:29 INFO mapreduce.Job: Running job: job_1557902299589_0011
20/06/11 15:12:35 INFO mapreduce.Job:Job job_1557902299589_0011 running in uber mode:false
20/06/11 15:12:35 INFO mapreduce.Job:map 0% reduce 0%
20/06/11 15:12:41 INFO mapreduce.Job:map 25% reduce 0%
20/06/11 15:12:42 INFO mapreduce.Job:map 50% reduce 0%
20/06/11 15:12:43 INFO mapreduce.Job:map 100% reduce 0%
20/06/11 15:12:44 INFO mapreduce.Job:Job job_1557902299589_0011 completed successfully
20/06/11 15:12:44 INFO mapreduce.Job:Counters:30
        File System Counters
                FILE:Number of bytes read=0
                FILE:Number of bytes written=494872
                FILE:Number of read operations=0
                FILE:Number of large read operations=0
                FILE:Number of write operations=0
                HDFS:Number of bytes read=511
                HDFS:Number of bytes written=9748
                HDFS:Number of read operations=16
                HDFS:Number of large read operations=0
                HDFS:Number of write operations=8
        Job Counters
                Launched map tasks=4
                Other local map tasks=4
                Total time spent by all maps in occupied slots (ms)=14102
```

```
                Total time spent by all reduces in occupied slots (ms) =0
                Total time spent by all map tasks (ms) =14102
                Total vcore - seconds taken by all map tasks =14102
                Total megabyte - seconds taken by all map tasks = 14440448
        Map - Reduce Framework
                Map input records =699
                Map output records =699
                Input split bytes =511
                Spilled Records =0
                Failed Shuffles =0
                Merged Map outputs =0
                GC time elapsed (ms) =398
                CPU time spent (ms) =8000
                Physical memory (bytes) snapshot =705671168
                Virtual memory (bytes) snapshot =8552480768
                Total committed heap usage (bytes) =573571072
        File Input Format Counters
                Bytes Read =0
        File Output Format Counters
                Bytes Written =9748
20/06/11 15:12:44 INFO mapreduce.ImportJobBase: Transferred 9.5195 KB in 17.9192 seconds (543.9975 bytes/sec)
20/06/11 15:12:44 INFO mapreduce.ImportJobBase: Retrieved 699 records.
```

执行完成之后，可以在 HDFS 上看到如图 6 - 5 所示的信息。

通过命令查看导入的文件：

```
[root@Superintendent2 ~]# hadoop fs - cat /user/YYR/my_help_keyword/part - m - 00000
0,(JSON)
1, ->
2, ->>
3, < >
4,ACCOUNT
5,ACTION
```

Hadoop Overview Datanodes Snapshot Startup Progress Utilities

Browse Directory

/user/master/my_help_keyword Go!

Permission	Owner	Group	Size	Replication	Block Size	Name
-rw-r--r--	root	supergroup	0 B	3	128 MB	_SUCCESS
-rw-r--r--	root	supergroup	1.98 KB	3	128 MB	part-m-00000
-rw-r--r--	root	supergroup	2.51 KB	3	128 MB	part-m-00001
-rw-r--r--	root	supergroup	2.3 KB	3	128 MB	part-m-00002
-rw-r--r--	root	supergroup	2.73 KB	3	128 MB	part-m-00003

Hadoop, 2014.

图6-5　普通导入操作完成

```
6,ADD
7,AES_DECRYPT
8,AES_ENCRYPT
9,AFTER
10,AGAINST
11,AGGREGATE
12,ALGORITHM
... (中间部分数据省略)
174,FETCH
[root@Superintendent2 ~]#
```

实例2：指定分隔符。

实例2：指定分隔符

```
# 数据导入命令
[root@Superintendent2 ~]# sqoop import --connect jdbc:mysql://YYR:3306/mysql --username root --password root00 --table help_keyword --target-dir /user/YYR/my_help_keyword1 --fields-terminated-by '\t' -m 2
# 数据导入命令执行过程如下
20/06/11 15:19:52 INFO sqoop.Sqoop:Running Sqoop version:1.4.7
20/06/11 15:19:52 WARN tool.BaseSqoopTool:Setting your password on the command-line is insecure. Consider using -P instead.
20/06/11 15:19:52 INFO manager.MySQLManager:Preparing to use a MySQL streaming resultset.
```

```
7. 20/06/11 15:19:52 INFO tool.CodeGenTool:Beginning code genera-
   tion
8.Loading class 'com.mysql.jdbc.Driver'. This is deprecated. The
   new driver class is 'com.mysql.cj.jdbc.Driver'. The driver is au-
   tomatically registered via the SPI and manual loading of the
   driver class is generally unnecessary.
9.20/06/11 15:19:53 INFO manager.SqlManager:Executing SQL state-
   ment:SELECT t.* FROM 'help_keyword' AS t LIMIT 1
10. 20/06/11 15:19:53 INFO manager.SqlManager:Executing SQL state-
    ment:SELECT t.* FROM 'help_keyword' AS t LIMIT 1
11.20/06/11 15:19:53 INFO orm.CompilationManager:HADOOP_MAPRED_
    HOME is/usr/local/hadoop
12. 注:/tmp/sqoop-root/compile/14cc2fd31c7a396d4ff258c6b862c9
    f1/help_keyword.java 使用或覆盖了已过时的 API。
13. 注:有关的详细信息,请使用 -Xlint:deprecation 重新编译。
14. 20/06/11 15:19:54 INFO orm.CompilationManager:Writing jar
    file:/tmp/sqoop-root/compile/14cc2fd31c7a396d4ff258c6b862
    c9f1/help_keyword.jar
15.20/06/11 15:19:54 WARN manager.MySQLManager:It looks like you
    are importing from mysql.
16.20/06/11 15:19:54 WARN manager.MySQLManager:This transfer can
    be faster! Use the --direct
17.20/06/11 15:19:54 WARN manager.MySQLManager:option to exercise
    a MySQL-specific fast path.
18.20/06/11 15:19:54 INFO manager.MySQLManager:Setting zero DATE-
    TIME behavior to convertToNull (mysql)
19.20/06/11 15:19:54 INFO mapreduce.ImportJobBase:Beginning im-
    port of help_keyword
20.20/06/11 15:19:54 INFO Configuration.deprecation:mapred.jar is
    deprecated. Instead,use mapreduce.job.jar
21.20/06/11 15:19:54 INFO Configuration.deprecation: mapred.
    map.tasks is deprecated. Instead,use mapreduce.job.maps
22.20/06/11 15:19:54 INFO client.RMProxy:Connecting to ResourceM-
    anager at YYR/192.168.3.190:8032
23.20/06/11 15:19:57 INFO db.DBInputFormat:Using read commited
    transaction isolation
```

```
20/06/11 15:19:57 INFO db.DataDrivenDBInputFormat: BoundingVals Query:SELECT MIN('help_keyword_id'),MAX('help_keyword_id') FROM 'help_keyword'
20/06/11 15:19:57 INFO db.IntegerSplitter:Split size:349;Num splits:2 from:0 to:698
20/06/11 15:19:57 INFO mapreduce.JobSubmitter: number of splits:2
20/06/11 15:19:57 INFO mapreduce.JobSubmitter:Submitting tokens for job:job_1557902299589_0012
20/06/11 15:19:57 INFO impl.YarnClientImpl:Submitted application application_1557902299589_0012
20/06/11 15:19:57 INFO mapreduce.Job:The url to track the job: http://YYR:8088/proxy/application_1557902299589_0012/
20/06/11 15:19:57 INFO mapreduce.Job: Running job: job_1557902299589_0012
20/06/11 15:20:03 INFO mapreduce.Job:Job job_1557902299589_0012 running in uber mode:false
20/06/11 15:20:03 INFO mapreduce.Job:map 0% reduce 0%
20/06/11 15:20:08 INFO mapreduce.Job:map 50% reduce 0%
20/06/11 15:20:09 INFO mapreduce.Job:map 100% reduce 0%
20/06/11 15:20:10 INFO mapreduce.Job:Job job_1557902299589_0012 completed successfully
20/06/11 15:20:10 INFO mapreduce.Job:Counters:30
        File System Counters
                FILE:Number of bytes read=0
                FILE:Number of bytes written=247438
                FILE:Number of read operations=0
                FILE:Number of large read operations=0
                FILE:Number of write operations=0
                HDFS:Number of bytes read=255
                HDFS:Number of bytes written=9748
                HDFS:Number of read operations=8
                HDFS:Number of large read operations=0
                HDFS:Number of write operations=4
        Job Counters
                Launched map tasks=2
                Other local map tasks=2
```

```
                Total time spent by all maps in occupied slots (ms)=6734
                Total time spent by all reduces in occupied slots (ms)=0
                Total time spent by all map tasks (ms)=6734
                Total vcore-seconds taken by all map tasks=6734
                Total megabyte-seconds taken by all map tasks=6895616
        Map-Reduce Framework
                Map input records=699
                Map output records=699
                Input split bytes=255
                Spilled Records=0
                Failed Shuffles=0
                Merged Map outputs=0
                GC time elapsed (ms)=224
                CPU time spent (ms)=4200
                Physical memory (bytes) snapshot=357216256
                Virtual memory (bytes) snapshot=4274778112
                Total committed heap usage (bytes)=277348352
        File Input Format Counters
                Bytes Read=0
        File Output Format Counters
                Bytes Written=9748
20/06/11 15:20:10 INFO mapreduce.ImportJobBase: Transferred 9.5195 KB in 16.0661 seconds (606.7416 bytes/sec)
20/06/11 15:20:10 INFO mapreduce.ImportJobBase: Retrieved 699 records.
```

执行完成之后，在 HDFS 中看到如图 6-6 所示的目录文件。

通过命令查看文件内容：

```
[root@Superintendent2 ~]# hdfs dfs -cat /user/YYR/my_help_keyword1/part-m-00000
0       (JSON)
1 ->
2 ->>
3       < >
```

Hadoop Overview Datanodes Snapshot Startup Progress Utilities

Browse Directory

/user/master/my_help_keyword1 Go!

Permission	Owner	Group	Size	Replication	Block Size	Name
-rw-r--r--	root	supergroup	0 B	3	128 MB	_SUCCESS
-rw-r--r--	root	supergroup	4.48 KB	3	128 MB	part-m-00000
-rw-r--r--	root	supergroup	5.04 KB	3	128 MB	part-m-00001

Hadoop, 2014.

图6-6　指定分隔符操作完成

```
4       ACCOUNT
5       ACTION
6       ADD
7       AES_DECRYPT
8       AES_ENCRYPT
9       AFTER
10      AGAINST
11      AGGREGATE
12      ALGORITHM
...  (中间部分数据省略)
348     MEDIUM
[root@Superintendent2 ~]#
```

实例3：带WHERE条件

实例3：带where条件。

```
# 数据导入命令
[root@Superintendent2 ~]# sqoop import --connect jdbc:mysql://YYR:3306/mysql --username root --password root00 --table help_keyword --target-dir/user/YYR/my_help_keyword2 --where "name='STRING'" -m 2        20/06/11 15:25:42
# 数据导入命令执行过程如下
INFO sqoop.Sqoop:Running Sqoop version:1.4.7
20/06/11 15:25:42 WARN tool.BaseSqoopTool:Setting your password on the command-line is insecure. Consider using -P instead.
20/06/11 15:25:42 INFO manager.MySQLManager:Preparing to use a MySQL streaming resultset.
```

```
7. 20/06/11 15:25:42 INFO tool.CodeGenTool:Beginning code genera-
   tion
8. Loading class 'com.mysql.jdbc.Driver'. This is deprecated. The
   new driver class is 'com.mysql.cj.jdbc.Driver'. The driver is au-
   tomatically registered via the SPI and manual loading of the
   driver class is generally unnecessary.
9. 20/06/11 15:25:43 INFO manager.SqlManager:Executing SQL state-
   ment:SELECT t.* FROM 'help_keyword' AS t LIMIT 1
10. 20/06/11 15:25:43 INFO manager.SqlManager:Executing SQL state-
   ment:SELECT t.* FROM 'help_keyword' AS t LIMIT 1
11. 20/06/11 15:25:43 INFO orm.CompilationManager:HADOOP_MAPRED_
   HOME is/usr/local/hadoop
12. 注:/tmp/sqoop-root/compile/691d61ee0180b4e99605f022af88900c/
   help_keyword.java 使用或覆盖了已过时的 API。
13. 注:有关的详细信息,请使用 -Xlint:deprecation 重新编译。
14. 20/06/11 15:25:44 INFO orm.CompilationManager:Writing jar
   file:/tmp/sqoop-root/compile/691d61ee0180b4e99605f022a
   f88900c/help_keyword.jar
15. 20/06/11 15:25:44 WARN manager.MySQLManager:It looks like you
   are importing from mysql.
16. 20/06/11 15:25:44 WARN manager.MySQLManager:This transfer can
   be faster! Use the --direct
17. 20/06/11 15:25:44 WARN manager.MySQLManager:option to exercise
   a MySQL-specific fast path.
18. 20/06/11 15:25:44 INFO manager.MySQLManager:Setting zero DATE-
   TIME behavior to convertToNull (mysql)
19. 20/06/11 15:25:44 INFO mapreduce.ImportJobBase:Beginning im-
   port of help_keyword
20. 20/06/11 15:25:44 INFO Configuration.deprecation:mapred.jar is
   deprecated. Instead,use mapreduce.job.jar
21. 20/06/11 15:25:44 INFO Configuration.deprecation:mapred.
   map.tasks is deprecated. Instead,use mapreduce.job.maps
22. 20/06/11 15:25:45 INFO client.RMProxy:Connecting to ResourceM-
   anager at YYR/192.168.3.190:8032
23. 20/06/11 15:25:46 INFO db.DBInputFormat:Using read commited
   transaction isolation
```

```
20/06/11 15:25:46 INFO db.DataDrivenDBInputFormat: BoundingValsQuery:SELECT MIN('help_keyword_id'),MAX('help_keyword_id') FROM 'help_keyword' WHERE ( name = 'STRING' )
20/06/11 15:25:46 INFO db.IntegerSplitter: Split size: 0; Num splits:2 from:553 to:553
20/06/11 15:25:47 INFO mapreduce.JobSubmitter: number of splits:1
20/06/11 15:25:47 INFO mapreduce.JobSubmitter: Submitting tokens for job:job_1557902299589_0013
20/06/11 15:25:47 INFO impl.YarnClientImpl: Submitted application application_1557902299589_0013
20/06/11 15:25:47 INFO mapreduce.Job: The url to track the job: http://YYR:8088/proxy/application_1557902299589_0013/
20/06/11 15:25:47 INFO mapreduce.Job: Running job: job_1557902299589_0013
20/06/11 15:25:52 INFO mapreduce.Job: Job job_1557902299589_0013 running in uber mode:false
20/06/11 15:25:52 INFO mapreduce.Job:map 0% reduce 0%
20/06/11 15:25:58 INFO mapreduce.Job:map 100% reduce 0%
20/06/11 15:25:58 INFO mapreduce.Job: Job job_1557902299589_0013 completed successfully
20/06/11 15:25:58 INFO mapreduce.Job:Counters:30
        File System Counters
                FILE:Number of bytes read=0
                FILE:Number of bytes written=123873
                FILE:Number of read operations=0
                FILE:Number of large read operations=0
                FILE:Number of write operations=0
                HDFS:Number of bytes read=129
                HDFS:Number of bytes written=11
                HDFS:Number of read operations=4
                HDFS:Number of large read operations=0
                HDFS:Number of write operations=2
        Job Counters
                Launched map tasks=1
                Other local map tasks=1
```

```
                Total time spent by all maps in occupied slots
                (ms)=2996
                Total time spent by all reduces in occupied slots
                (ms)=0
                Total time spent by all map tasks (ms)=2996
                Total vcore-seconds taken by all map tasks=2996
                Total megabyte-seconds taken by all map tasks=
                3067904
        Map-Reduce Framework
                Map input records=1
                Map output records=1
                Input split bytes=129
                Spilled Records=0
                Failed Shuffles=0
                Merged Map outputs=0
                GC time elapsed (ms)=100
                CPU time spent (ms)=1990
                Physical memory (bytes) snapshot=179253248
                Virtual memory (bytes) snapshot=2137178112
                Total committed heap usage (bytes)=134742016
        File Input Format Counters
                Bytes Read=0
        File Output Format Counters
                Bytes Written=11
20/06/11 15:25:58 INFO mapreduce.ImportJobBase:Transferred 11
  bytes in 13.7931 seconds (0.7975 bytes/sec)
20/06/11 15:25:58 INFO mapreduce.ImportJobBase:Retrieved 1 re-
  cords.
```

执行完成之后，在 HDFS 中看到如图 6-7 所示的目录文件。

通过命令查看文件内容：

```
[root@Superintendent2 ~]# hdfs dfs -cat /user/YYR/my_help_key-
  word2/part-m-00000
553,STRING
[root@Superintendent2 ~]#
```

实例 4：导入指定列。

实例 4：导入指定列

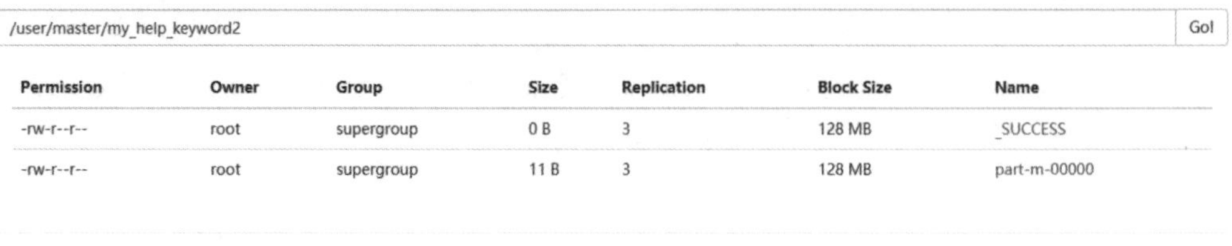

Permission	Owner	Group	Size	Replication	Block Size	Name
-rw-r--r--	root	supergroup	0 B	3	128 MB	_SUCCESS
-rw-r--r--	root	supergroup	11 B	3	128 MB	part-m-00000

图 6-7　带 where 条件操作完成

```
1.# 数据导入命令
2.[root@Superintendent2 ~]# sqoop import --connect jdbc:mysql://
  YYR:3306/mysql --username root --password root00 --table help_
  keyword --target-dir/user/YYR/my_help_keyword3 --where "name
  ='LONG'" --columns "name" -m 2
3.# 数据导入命令执行过程如下
4.20/06/11 15:31:02 INFO sqoop.Sqoop:Running Sqoop version:1.4.7
5.20/06/11 15:31:02 WARN tool.BaseSqoopTool:Setting your password
  on the command-line is insecure. Consider using -P instead.
6.20/06/11 15:31:02 INFO manager.MySQLManager:Preparing to use a
  MySQL streaming resultset.
7.20/06/11 15:31:02 INFO tool.CodeGenTool:Beginning code genera-
  tion
8.Loading class 'com.mysql.jdbc.Driver'. This is deprecated. The
  new driver class is 'com.mysql.cj.jdbc.Driver'. The driver is
  automatically registered via the SPI and manual loading of the
  driver class is generally unnecessary.
9.20/06/11 15:31:03 INFO manager.SqlManager:Executing SQL state-
  ment:SELECT t.* FROM 'help_keyword' AS t LIMIT 1
10.20/06/11 15:31:03 INFO manager.SqlManager:Executing SQL state-
   ment:SELECT t.* FROM 'help_keyword' AS t LIMIT 1
11.20/06/11 15:31:03 INFO orm.CompilationManager:HADOOP_MAPRED_
   HOME is/usr/local/hadoop
12.注:/tmp/sqoop-root/compile/bc192659eb79c3f9b27aca7ac223e377/
   help_keyword.java 使用或覆盖了已过时的 API。
13.注:有关的详细信息,请使用 -Xlint:deprecation 重新编译。
```

14. 20/06/11 15:31:04 INFO orm.CompilationManager:Writing jar file:/tmp/sqoop-root/compile/bc192659eb79c3f9b27aca7ac223e377/help_keyword.jar
15. 20/06/11 15:31:04 WARN manager.MySQLManager:It looks like you are importing from mysql.
16. 20/06/11 15:31:04 WARN manager.MySQLManager:This transfer can be faster! Use the --direct
17. 20/06/11 15:31:04 WARN manager.MySQLManager:option to exercise a MySQL-specific fast path.
18. 20/06/11 15:31:04 INFO manager.MySQLManager:Setting zero DATETIME behavior to convertToNull (mysql)
19. 20/06/11 15:31:04 INFO mapreduce.ImportJobBase:Beginning import of help_keyword
20. 20/06/11 15:31:04 INFO Configuration.deprecation:mapred.jar is deprecated. Instead,use mapreduce.job.jar
21. 20/06/11 15:31:05 INFO Configuration.deprecation:mapred.map.tasks is deprecated. Instead,use mapreduce.job.maps
22. 20/06/11 15:31:05 INFO client.RMProxy:Connecting to ResourceManager at YYR/192.168.3.190:8032
23. 20/06/11 15:31:07 INFO db.DBInputFormat:Using read commited transaction isolation
24. 20/06/11 15:31:07 INFO db.DataDrivenDBInputFormat:BoundingValsQuery:SELECT MIN('help_keyword_id'),MAX('help_keyword_id') FROM 'help_keyword' WHERE (name = 'LONG')
25. 20/06/11 15:31:07 INFO db.IntegerSplitter:Split size:0;Num splits:2 from:309 to:309
26. 20/06/11 15:31:07 INFO mapreduce.JobSubmitter:number of splits:1
27. 20/06/11 15:31:07 INFO mapreduce.JobSubmitter:Submitting tokens for job:job_1557902299589_0014
28. 20/06/11 15:31:07 INFO impl.YarnClientImpl:Submitted application application_1557902299589_0014
29. 20/06/11 15:31:07 INFO mapreduce.Job:The url to track the job:http://YYR:8088/proxy/application_1557902299589_0014/
30. 20/06/11 15:31:07 INFO mapreduce.Job:Running job:job_1557902299589_0014

```
20/06/11 15:31:12 INFO mapreduce.Job:Job job_1557902299589_0014 running in uber mode:false
20/06/11 15:31:12 INFO mapreduce.Job:map 0% reduce 0%
20/06/11 15:31:18 INFO mapreduce.Job:map 100% reduce 0%
20/06/11 15:31:18 INFO mapreduce.Job:Job job_1557902299589_0014 completed successfully
20/06/11 15:31:18 INFO mapreduce.Job:Counters:30
        File System Counters
                FILE:Number of bytes read=0
                FILE:Number of bytes written=123853
                FILE:Number of read operations=0
                FILE:Number of large read operations=0
                FILE:Number of write operations=0
                HDFS:Number of bytes read=129
                HDFS:Number of bytes written=5
                HDFS:Number of read operations=4
                HDFS:Number of large read operations=0
                HDFS:Number of write operations=2
        Job Counters
                Launched map tasks=1
                Other local map tasks=1
                Total time spent by all maps in occupied slots (ms)=3008
                Total time spent by all reduces in occupied slots (ms)=0
                Total time spent by all map tasks (ms)=3008
                Total vcore-seconds taken by all map tasks=3008
                Total megabyte-seconds taken by all map tasks=3080192
        Map-Reduce Framework
                Map input records=1
                Map output records=1
                Input split bytes=129
                Spilled Records=0
                Failed Shuffles=0
                Merged Map outputs=0
                GC time elapsed (ms)=85
```

```
                CPU time spent (ms) =1900
                Physical memory (bytes) snapshot =179470336
                Virtual memory (bytes) snapshot =2135539712
                Total committed heap usage (bytes) =149946368
        File Input Format Counters
                Bytes Read =0
        File Output Format Counters
                Bytes Written =5
20/06/11 15:31:18 INFO mapreduce.ImportJobBase:Transferred 5 bytes in 13.4994 seconds (0.3704 bytes/sec)
20/06/11 15:31:18 INFO mapreduce.ImportJobBase:Retrieved 1 records.
```

执行完成之后，在 HDFS 中看到如图 6－8 所示的目录文件。

Hadoop Overview Datanodes Snapshot Startup Progress Utilities

Browse Directory

/user/master/my_help_keyword3 Go!

Permission	Owner	Group	Size	Replication	Block Size	Name
-rw-r--r--	root	supergroup	0 B	3	128 MB	_SUCCESS
-rw-r--r--	root	supergroup	5 B	3	128 MB	part-m-00000

Hadoop, 2014.

图 6－8 导入指定列操作完成

通过命令查看文件内容：

```
[root@Superintendent2 ~]# hdfs dfs -cat /user/YYR/my_help_keyword3/part-m-00000
LONG
[root@Superintendent2 ~]#
```

实例 5：自定义 SQL

实例 5：自定义 SQL。

```
# 数据导入命令
[root@Superintendent2 ~]# sqoop import --connect jdbc:mysql://YYR:3306/mysql -username root --password root00 --target-dir/user/YYR/my_help_keyword4 --query 'select name,help_keyword_id from help_keyword where $CONDITIONS and name = "STRING"' --split-by help_keyword_id -m 2
```

```
3.#数据导入命令执行过程如下
4.20/06/11 15:38:33 INFO sqoop.Sqoop:Running Sqoop version:1.4.7
5. 20/06/11 15:38:33 WARN tool.BaseSqoopTool:Setting your password
   on the command-line is insecure. Consider using -P instead.
6.20/06/11 15:38:33 INFO manager.MySQLManager:Preparing to use a
   MySQL streaming resultset.
7.20/06/11 15:38:33 INFO tool.CodeGenTool:Beginning code genera-
   tion
8.Loading class 'com.mysql.jdbc.Driver'. This is deprecated. The
   new driver class is 'com.mysql.cj.jdbc.Driver'. The driver is
   automatically registered via the SPI and manual loading of the
   driver class is generally unnecessary.
9.20/06/11 15:38:34 INFO manager.SqlManager:Executing SQL state-
   ment:select name,help_keyword_id from help_keyword where  (1=
   0)  and name="STRING"
10. 20/06/11 15:38:34 INFO manager.SqlManager:Executing SQL state-
    ment:select name,help_keyword_id from help_keyword where  (1=
    0)  and name="STRING"
11.20/06/11 15:38:34 INFO manager.SqlManager:Executing SQL state-
    ment:select name,help_keyword_id from help_keyword where  (1=
    0)  and name="STRING"
12.20/06/11 15:38:34 INFO orm.CompilationManager:HADOOP_MAPRED_
    HOME is/usr/local/hadoop
13. 注:/tmp/sqoop-root/compile/df3d5720bc0d0eb7c2a7d026827a06d1/
    QueryResult.java 使用或覆盖了已过时的 API。
14. 注:有关的详细信息,请使用 -Xlint:deprecation 重新编译。
15.20/06/11 15:38:35 INFO orm.CompilationManager:Writing jar
    file:/tmp/sqoop-root/compile/df3d5720bc0d0eb7c2a7d026827a06d1/
    QueryResult.jar
16.20/06/11 15:38:35 INFO mapreduce.ImportJobBase:Beginning query
    import.
17.20/06/11 15:38:35 INFO Configuration.deprecation:mapred.jar is
    deprecated. Instead,use mapreduce.job.jar
18.20/06/11 15:38:36 INFO Configuration.deprecation:mapred.map.
    tasks is deprecated. Instead,use mapreduce.job.maps
19.20/06/11 15:38:36 INFO client.RMProxy:Connecting to Resource
    Manager at YYR/192.168.3.190:8032
```

```
20/06/11 15:38:37 INFO db.DBInputFormat:Using read commited transaction isolation
20/06/11 15:38:37 INFO db.DataDrivenDBInputFormat: BoundingValsQuery:SELECT MIN(help_keyword_id),MAX(help_keyword_id) FROM (select name,help_keyword_id from help_keyword where (1 =1)  and name = "STRING") AS t1
20/06/11 15:38:37 INFO db.IntegerSplitter: Split size:0;Num splits:2 from:553 to:553
20/06/11 15:38:37 INFO mapreduce.JobSubmitter: number of splits:1
20/06/11 15:38:38 INFO mapreduce.JobSubmitter:Submitting tokens for job:job_1557902299589_0015
20/06/11 15:38:38 INFO impl.YarnClientImpl:Submitted application application_1557902299589_0015
20/06/11 15:38:38 INFO mapreduce.Job:The url to track the job: http://YYR:8088/proxy/application_1557902299589_0015/
20/06/11 15:38:38 INFO mapreduce.Job: Running job: job_1557902299589_0015
20/06/11 15:38:44 INFO mapreduce.Job:Job job_1557902299589_0015 running in uber mode:false
20/06/11 15:38:44 INFO mapreduce.Job:map 0%  reduce 0%
20/06/11 15:38:49 INFO mapreduce.Job:map 100%  reduce 0%
20/06/11 15:38:49 INFO mapreduce.Job:Job job_1557902299589_0015 completed successfully
20/06/11 15:38:49 INFO mapreduce.Job:Counters:30
        File System Counters
                FILE:Number of bytes read =0
                FILE:Number of bytes written =123866
                FILE:Number of read operations =0
                FILE:Number of large read operations =0
                FILE:Number of write operations =0
                HDFS:Number of bytes read =125
                HDFS:Number of bytes written =11
                HDFS:Number of read operations =4
                HDFS:Number of large read operations =0
                HDFS:Number of write operations =2
        Job Counters
```

```
                Launched map tasks =1
                Other local map tasks =1
                Total time spent by all maps in occupied slots (ms) =2877
                Total time spent by all reduces in occupied slots (ms) =0
                Total time spent by all map tasks (ms) =2877
                Total vcore - seconds taken by all map tasks =2877
                Total megabyte - seconds taken by all map tasks = 2946048
        Map - Reduce Framework
                Map input records =1
                Map output records =1
                Input split bytes =125
                Spilled Records =0
                Failed Shuffles =0
                Merged Map outputs =0
                GC time elapsed (ms) =85
                CPU time spent (ms) =1860
                Physical memory (bytes) snapshot =176635904
                Virtual memory (bytes) snapshot =2135904256
                Total committed heap usage (bytes) =152043520
        File Input Format Counters
                Bytes Read =0
        File Output Format Counters
                Bytes Written =11
20/06/11 15:38:49 INFO mapreduce.ImportJobBase:Transferred 11 bytes in 13.6189 seconds (0.8077 bytes/sec)
20/06/11 15:38:49 INFO mapreduce.ImportJobBase:Retrieved 1 records.
```

执行完成之后，在 HDFS 中看到如图 6-9 所示的目录文件。

通过命令查看文件内容：

```
[root@Superintendent2 ~]# hdfs dfs -cat /user/YYR/my_help_keyword4/part-m-00000
STRING,553
[root@Superintendent2 ~]#
```

在以上需要按照自定义 SQL 语句导出数据到 HDFS 的情况下，需注意：

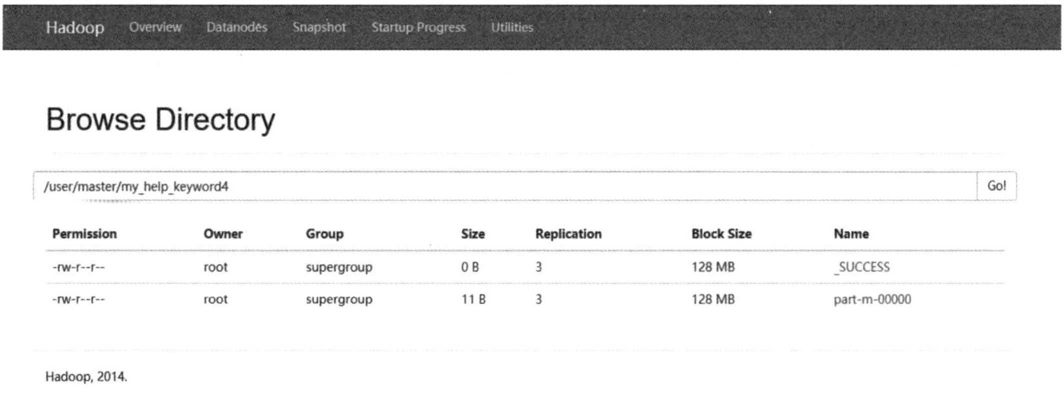

Hadoop Overview Datanodes Snapshot Startup Progress Utilities

Browse Directory

/user/master/my_help_keyword4 Go!

Permission	Owner	Group	Size	Replication	Block Size	Name
-rw-r--r--	root	supergroup	0 B	3	128 MB	_SUCCESS
-rw-r--r--	root	supergroup	11 B	3	128 MB	part-m-00000

Hadoop, 2014.

图 6-9 自定义 SQL 操作完成

①引号问题，要么外层使用单引号，内层使用双引号，$CONDITIONS 的 $ 符号不用转义；要么外层使用双引号，内层使用单引号，$CONDITIONS 的 $ 符号需要转义。

②自定义的 SQL 语句中必须带有 WHERE $CONDITIONS。

- 把 MySQL 数据库中的表数据导入到 Hive 中

要把 MySQL 数据库中的表数据导入到 Hive 中，首先，将相应的表数据导入到 HDFS 中，然后把表数据类型映射为 Hive 表数据类型。根据表结构，在 Hive 上执行 create table 操作创建 Hive 表。最后，在 Hive 中执行 load data input 语句，将 HDFS 上的表数据移动到 Hive 数据仓库目录。

导入过程：

第一步：导入 mysql. help_keyword 的数据到 HDFS 的默认路径。

第二步：自动仿造 mysql. help_keyword 去创建一张 Hive 表，创建在默认的 default 库中。

第三步：把临时目录中的数据导入到 Hive 表中。

为了能够顺利地完成 Hive 数据的导入，需要先在 Hive 中创建数据库。

```
[root@Superintendent2 ~]# hive

Logging initialized using configuration in jar:file:/usr/local/hive/lib/hive-common-1.1.0.jar!/hive-log4j.properties
SLF4J:Class path contains multiple SLF4J bindings.
SLF4J:Found binding in[jar:file:/usr/local/hadoop/share/hadoop/common/lib/slf4j-log4j12-1.7.5.jar!/org/slf4j/impl/StaticLoggerBinder.class]
SLF4J:Found binding in[jar:file:/usr/local/hive/lib/hive-jdbc-1.1.0-standalone.jar!/org/slf4j/impl/StaticLoggerBinder.class]
```

```
SLF4J:See http://www.slf4j.org/codes.html#multiple_bindings for an explanation.
SLF4J:Actual binding is of type[org.slf4j.impl.Log4jLoggerFactory]
#创建my_keyword数据库
hive>create database my_keyword;
OK
Time taken:0.454 seconds
hive>
```

实例1：普通导入。

```
#数据导入命令
[root@Superintendent2 ~]# sqoop import --connect jdbc:mysql://YYR:3306/mysql --username root --password root00 --table help_keyword --hive-import -m 1 --hive-database my_keyword --hive-table my_keyword --create-hive-table --hive-overwrite
#数据导入命令执行过程如下
20/06/11 15:59:19 INFO sqoop.Sqoop:Running Sqoop version:1.4.7
20/06/11 15:59:19 WARN tool.BaseSqoopTool:Setting your password on the command-line is insecure. Consider using -P instead.
20/06/11 15:59:19 INFO tool.BaseSqoopTool:Using Hive-specific delimiters for output. You can override
20/06/11 15:59:19 INFO tool.BaseSqoopTool:delimiters with --fields-terminated-by,etc.
20/06/11 15:59:19 INFO manager.MySQLManager:Preparing to use a MySQL streaming resultset.
20/06/11 15:59:19 INFO tool.CodeGenTool:Beginning code generation
Loading class 'com.mysql.jdbc.Driver'. This is deprecated. The new driver class is 'com.mysql.cj.jdbc.Driver'. The driver is automatically registered via the SPI and manual loading of the driver class is generally unnecessary.
20/06/11 15:59:20 INFO manager.SqlManager:Executing SQL statement:SELECT t.* FROM 'help_keyword' AS t LIMIT 1
20/06/11 15:59:20 INFO manager.SqlManager:Executing SQL statement:SELECT t.* FROM 'help_keyword' AS t LIMIT 1
```

```
13. 20/06/11 15:59:20 INFO orm.CompilationManager:HADOOP_MAPRED_
    HOME is/usr/local/hadoop
14. 注:/tmp/sqoop-root/compile/3ab5170027d4b184aa6b49a9e751e6be/
    help_keyword. java 使用或覆盖了已过时的 API。
15. 注:有关的详细信息,请使用 -Xlint:deprecation 重新编译。
16. 20/06/11 15:59:21 INFO orm.CompilationManager: Writing jar
    file:/tmp/sqoop-root/compile/3ab5170027d4b184aa6b49a9e751e6be/
    help_keyword. jar
17.20/06/11 15:59:21 WARN manager.MySQLManager:It looks like you
    are importing from mysql.
18.20/06/11 15:59:21 WARN manager.MySQLManager:This transfer can
    be faster! Use the --direct
19.20/06/11 15:59:21 WARN manager.MySQLManager:option to exercise
    a MySQL-specific fast path.
20.20/06/11 15:59:21 INFO manager.MySQLManager:Setting zero DATE-
    TIME behavior to convertToNull (mysql)
21.20/06/11 15:59:21 INFO mapreduce.ImportJobBase:Beginning im-
    port of help_keyword
22.20/06/11 15:59:21 INFO Configuration.deprecation:mapred.jar is
    deprecated. Instead,use mapreduce.job.jar
23.20/06/11 15:59:22 INFO Configuration.deprecation:mapred.map.
    tasks is deprecated. Instead,use mapreduce.job.maps
24.20/06/11 15:59:22 INFO client.RMProxy:Connecting to Resource
    Manager at YYR/192.168.3.190:8032
25.20/06/11 15:59:24 INFO db.DBInputFormat:Using read commited
    transaction isolation
26.20/06/11 15:59:24 INFO mapreduce.JobSubmitter: number of
    splits:1
27.20/06/11 15:59:24 INFO mapreduce.JobSubmitter:Submitting to-
    kens for job:job_1557902299589_0018
28.20/06/11 15:59:24 INFO impl.YarnClientImpl:Submitted applica-
    tion application_1557902299589_0018
29.20/06/11 15:59:24 INFO mapreduce.Job:The url to track the job:
    http://YYR:8088/proxy/application_1557902299589_0018/
30.20/06/11 15:59:24 INFO mapreduce.Job:Running job:job_1557902299589_
    0018
```

```
20/06/11 15:59:30 INFO mapreduce.Job:Job job_1557902299589_0018 running in uber mode:false
20/06/11 15:59:30 INFO mapreduce.Job:map 0%  reduce 0%
20/06/11 15:59:35 INFO mapreduce.Job:map 100%  reduce 0%
20/06/11 15:59:35 INFO mapreduce.Job:Job job_1557902299589_0018 completed successfully
20/06/11 15:59:36 INFO mapreduce.Job:Counters:30
        File System Counters
                FILE:Number of bytes read=0
                FILE:Number of bytes written=124139
                FILE:Number of read operations=0
                FILE:Number of large read operations=0
                FILE:Number of write operations=0
                HDFS:Number of bytes read=87
                HDFS:Number of bytes written=9748
                HDFS:Number of read operations=4
                HDFS:Number of large read operations=0
                HDFS:Number of write operations=2
        Job Counters
                Launched map tasks=1
                Other local map tasks=1
                Total time spent by all maps in occupied slots (ms)=2892
                Total time spent by all reduces in occupied slots (ms)=0
                Total time spent by all map tasks (ms)=2892
                Total vcore-seconds taken by all map tasks=2892
                Total megabyte-seconds taken by all map tasks=2961408
        Map-Reduce Framework
                Map input records=699
                Map output records=699
                Input split bytes=87
                Spilled Records=0
                Failed Shuffles=0
                Merged Map outputs=0
                GC time elapsed (ms)=64
```

```
                CPU time spent (ms)=1920
                Physical memory (bytes) snapshot=178565120
                Virtual memory (bytes) snapshot=2141843456
                Total committed heap usage (bytes)=144703488
        File Input Format Counters
                Bytes Read=0
        File Output Format Counters
                Bytes Written=9748
20/06/11 15:59:36 INFO mapreduce.ImportJobBase:Transferred 9.5195 KB in 13.6362 seconds (714.8641 bytes/sec)
20/06/11 15:59:36 INFO mapreduce.ImportJobBase:Retrieved 699 records.
20/06/11 15:59:36 INFO mapreduce.ImportJobBase: Publishing Hive/Hcat import job data to Listeners for table help_keyword
20/06/11 15:59:36 INFO manager.SqlManager:Executing SQL statement:SELECT t.* FROM 'help_keyword' AS t LIMIT 1
20/06/11 15:59:36 INFO hive.HiveImport:Loading uploaded data into Hive
20/06/11 15:59:37 INFO hive.HiveImport:
20/06/11 15:59:37 INFO hive.HiveImport: Logging initialized using configuration in jar:file:/usr/local/sqoop/lib/hive-exec-1.1.0.jar!/hive-log4j.properties
20/06/11 15:59:37 INFO hive.HiveImport:SLF4J:Class path contains multiple SLF4J bindings.
20/06/11 15:59:37 INFO hive.HiveImport:SLF4J:Found binding in [jar:file:/usr/local/hadoop/share/hadoop/common/lib/slf4j-log4j12-1.7.5.jar!/org/slf4j/impl/StaticLoggerBinder.class]
20/06/11 15:59:37 INFO hive.HiveImport:SLF4J:Found binding in [jar:file:/usr/local/hive/lib/hive-jdbc-1.1.0-standalone.jar!/org/slf4j/impl/StaticLoggerBinder.class]
20/06/11 15:59:37 INFO hive.HiveImport:SLF4J:See http://www.slf4j.org/codes.html#multiple_bindings for an explanation.
20/06/11 15:59:37 INFO hive.HiveImport:SLF4J:Actual binding is of type[org.slf4j.impl.Log4jLoggerFactory]
20/06/11 15:59:39 INFO hive.HiveImport:OK
```

```
20/06/11 15:59:39 INFO hive.HiveImport:Time taken:0.86 seconds
20/06/11 15:59:39 INFO hive.HiveImport:Loading data to table my_keyword.my_keyword
20/06/11 15:59:40 INFO hive.HiveImport:Table my_keyword.my_keyword stats:[numFiles=1,numRows=0,totalSize=9748,rawDataSize=0]
20/06/11 15:59:40 INFO hive.HiveImport:OK
20/06/11 15:59:40 INFO hive.HiveImport:Time taken:0.661 seconds
20/06/11 15:59:40 INFO hive.HiveImport:Hive import complete.
20/06/11 15:59:40 INFO hive.HiveImport:Export directory is contains the _SUCCESS file only,removing the directory.
```

执行完成之后，在 HDFS 中看到如图 6-10 所示的目录文件。

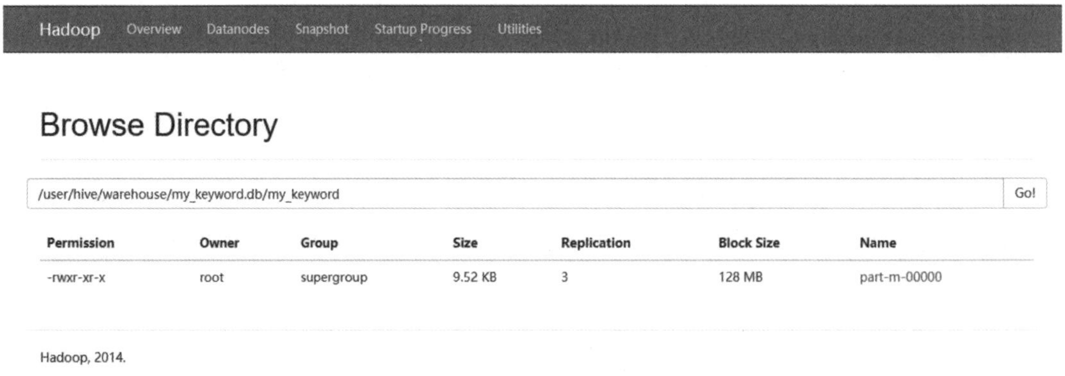

Hadoop Overview Datanodes Snapshot Startup Progress Utilities

Browse Directory

/user/hive/warehouse/my_keyword.db/my_keyword Go!

Permission	Owner	Group	Size	Replication	Block Size	Name
-rwxr-xr-x	root	supergroup	9.52 KB	3	128 MB	part-m-00000

Hadoop, 2014.

图 6-10 普通导入操作完成

通过命令查看 HDFS 文件数据如下：

```
[root@Superintendent2 ~]# hdfs dfs -cat /user/hive/warehouse/my_keyword.db/my_keyword/part-m-00000
0(JSON)
1 ->
2 ->>
3 < >
4ACCOUNT
5ACTION
6ADD
...  (中间部分省略)
698ZEROFILL
```

通过 Hive 命令查看 Hive 表数据如下：

```
[root@Superintendent2 ~]# hive
Logging initialized using configuration in jar:file:/usr/local/hive/lib/hive-common-1.1.0.jar!/hive-log4j.properties
SLF4J:Class path contains multiple SLF4J bindings.
SLF4J:Found binding in[jar:file:/usr/local/hadoop/share/hadoop/common/lib/slf4j-log4j12-1.7.5.jar!/org/slf4j/impl/StaticLoggerBinder.class]
SLF4J:Found binding in[jar:file:/usr/local/hive/lib/hive-jdbc-1.1.0-standalone.jar!/org/slf4j/impl/StaticLoggerBinder.class]
SLF4J:See http://www.slf4j.org/codes.html#multiple_bindings for an explanation.
SLF4J:Actual binding is of type[org.slf4j.impl.Log4jLoggerFactory]
hive >use my_keyword;
OK
Time taken:0.454 seconds
hive >show tables;
OK
my_keyword
Time taken:0.15 seconds,Fetched:1 row(s)
hive >select * from my_keyword;
OK
0       (JSON)
1 ->
2 ->>
3       < >
4       ACCOUNT
5       ACTION
6       ADD
7       AES_DECRYPT
8       AES_ENCRYPT
9       AFTER
  ...  (中间部分省略)
698     ZEROFILL
```

```
30.Time taken:0.354 seconds,Fetched:699 row(s)
31.hive >
```

实例2：指定行分隔符和列分隔符。

-- hive - overwrite 指定覆盖导入，-- create - hive - table 指定自动创建 Hive 表，-- hive - table 指定表名，-- delete - target - dir 指定删除中间结果数据目录，-- hive - database 指定数据库。

```
1.#数据导入命令
2.[root@Superintendent2 ~]# sqoop import --connect jdbc:mysql://YYR:3306/mysql --username root --password root00 --table help_keyword --hive-import -m 1 --hive-database my_keyword --hive-table my_keyword1 --create-hive-table --hive-overwrite --fields-terminated-by ',' --lines-terminated-by "\n"
3.#数据导入命令执行过程如下
4.20/06/11 16:07:05 INFO sqoop.Sqoop:Running Sqoop version:1.4.7
5.20/06/11 16:07:05 WARN tool.BaseSqoopTool:Setting your password on the command-line is insecure. Consider using -P instead.
6.20/06/11 16:07:05 INFO manager.MySQLManager:Preparing to use a MySQL streaming resultset.
7.20/06/11 16:07:05 INFO tool.CodeGenTool:Beginning code generation
8.Loading class 'com.mysql.jdbc.Driver'. This is deprecated. The new driver class is 'com.mysql.cj.jdbc.Driver'. The driver is automatically registered via the SPI and manual loading of the driver class is generally unnecessary.
9.20/06/11 16:07:06 INFO manager.SqlManager:Executing SQL statement:SELECT t.* FROM 'help_keyword' AS t LIMIT 1
10.20/06/11 16:07:06 INFO manager.SqlManager:Executing SQL statement:SELECT t.* FROM 'help_keyword' AS t LIMIT 1
11.20/06/11 16:07:06 INFO orm.CompilationManager:HADOOP_MAPRED_HOME is/usr/local/hadoop
12.注:/tmp/sqoop-root/compile/aed10b78d7be90ce82c4f90cbc7f399d/help_keyword.java 使用或覆盖了已过时的 API。
13.注:有关的详细信息,请使用 -Xlint:deprecation 重新编译。
14.20/06/11 16:07:07 INFO orm.CompilationManager:Writing jar file:/tmp/sqoop-root/compile/aed10b78d7be90ce82c4f90cbc7f399d/help_keyword.jar
```

```
20/06/11 16:07:07 WARN manager.MySQLManager:It looks like you are importing from mysql.
20/06/11 16:07:07 WARN manager.MySQLManager:This transfer can be faster! Use the --direct
20/06/11 16:07:07 WARN manager.MySQLManager:option to exercise a MySQL-specific fast path.
20/06/11 16:07:07 INFO manager.MySQLManager:Setting zero DATETIME behavior to convertToNull (mysql)
20/06/11 16:07:07 INFO mapreduce.ImportJobBase:Beginning import of help_keyword
20/06/11 16:07:07 INFO Configuration.deprecation:mapred.jar is deprecated. Instead,use mapreduce.job.jar
20/06/11 16:07:08 INFO Configuration.deprecation:mapred.map.tasks is deprecated. Instead,use mapreduce.job.maps
20/06/11 16:07:08 INFO client.RMProxy:Connecting to ResourceManager at YYR/192.168.3.190:8032
20/06/11 16:07:09 INFO db.DBInputFormat:Using read commited transaction isolation
20/06/11 16:07:10 INFO mapreduce.JobSubmitter: number of splits:1
20/06/11 16:07:10 INFO mapreduce.JobSubmitter:Submitting tokens for job:job_1557902299589_0019
20/06/11 16:07:10 INFO impl.YarnClientImpl:Submitted application application_1557902299589_0019
20/06/11 16:07:10 INFO mapreduce.Job:The url to track the job:http://YYR:8088/proxy/application_1557902299589_0019/
20/06/11 16:07:10 INFO mapreduce.Job:Running job:job_1557902299589_0019
20/06/11 16:07:16 INFO mapreduce.Job:Job job_1557902299589_0019 running in uber mode:false
20/06/11 16:07:16 INFO mapreduce.Job:map 0% reduce 0%
20/06/11 16:07:21 INFO mapreduce.Job:map 100% reduce 0%
20/06/11 16:07:22 INFO mapreduce.Job:Job job_1557902299589_0019 completed successfully
20/06/11 16:07:22 INFO mapreduce.Job:Counters:30
        File System Counters
                FILE:Number of bytes read=0
```

```
                FILE:Number of bytes written=124139
                FILE:Number of read operations=0
                FILE:Number of large read operations=0
                FILE:Number of write operations=0
                HDFS:Number of bytes read=87
                HDFS:Number of bytes written=9748
                HDFS:Number of read operations=4
                HDFS:Number of large read operations=0
                HDFS:Number of write operations=2
        Job Counters
                Launched map tasks=1
                Other local map tasks=1
                Total time spent by all maps in occupied slots (ms)=3363
                Total time spent by all reduces in occupied slots (ms)=0
                Total time spent by all map tasks (ms)=3363
                Total vcore-seconds taken by all map tasks=3363
                Total megabyte-seconds taken by all map tasks=3443712
        Map-Reduce Framework
                Map input records=699
                Map output records=699
                Input split bytes=87
                Spilled Records=0
                Failed Shuffles=0
                Merged Map outputs=0
                GC time elapsed (ms)=124
                CPU time spent (ms)=2000
                Physical memory (bytes) snapshot=179240960
                Virtual memory (bytes) snapshot=2139471872
                Total committed heap usage (bytes)=163053568
        File Input Format Counters
                Bytes Read=0
        File Output Format Counters
                Bytes Written=9748
```

```
69. 20/06/11 16:07:22 INFO mapreduce.ImportJobBase:Transferred
    9.5195 KB in 14.5204 seconds (671.33 bytes/sec)
70.20/06/11 16:07:22 INFO mapreduce.ImportJobBase:Retrieved 699
    records.
71.20/06/11 16:07:22 INFO mapreduce.ImportJobBase: Publishing
    Hive/Hcat import job data to Listeners for table help_keyword
72.20/06/11 16:07:22 INFO manager.SqlManager:Executing SQL state-
    ment:SELECT t. * FROM 'help_keyword' AS t LIMIT 1
73. 20/06/11 16:07:22 INFO hive.HiveImport:Loading uploaded data
    into Hive
74.20/06/11 16:07:24 INFO hive.HiveImport:
75. 20/06/11 16:07:24 INFO hive.HiveImport:Logging initialized
    using configuration in jar:file:/usr/local/sqoop/lib/hive-
    exec-1.1.0.jar!/hive-log4j.properties
76.20/06/11 16:07:24 INFO hive.HiveImport:SLF4J:Class path con-
    tains multiple SLF4J bindings.
77.20/06/11 16:07:24 INFO hive.HiveImport:SLF4J:Found binding in
    [jar:file:/usr/local/hadoop/share/hadoop/common/lib/slf4j-
    log4j12-1.7.5.jar!/org/slf4j/impl/StaticLoggerBinder.
    class]
78.20/06/11 16:07:24 INFO hive.HiveImport:SLF4J:Found binding in
    [jar:file:/usr/local/hive/lib/hive-jdbc-1.1.0-stand
    alone.jar!/org/slf4j/impl/StaticLoggerBinder.class]
79.20/06/11 16:07:24 INFO hive.HiveImport:SLF4J:See http://
    www.slf4j.org/codes.html#multiple_bindings for an explana-
    tion.
80.20/06/11 16:07:24 INFO hive.HiveImport:SLF4J:Actual binding is
    of type[org.slf4j.impl.Log4jLoggerFactory]
81.20/06/11 16:07:26 INFO hive.HiveImport:OK
82. 20/06/11 16:07:26 INFO hive.HiveImport: Time taken:
    0.996 seconds
83. 20/06/11 16:07:26 INFO hive.HiveImport:Loading data to table
    my_keyword.my_keyword1
84.20/06/11 16:07:27 INFO hive.HiveImport:Table my_keyword.my_
    keyword1 stats:[numFiles=1,numRows=0,totalSize=9748,raw-
    DataSize=0]
85.20/06/11 16:07:27 INFO hive.HiveImport:OK
```

```
20/06/11 16:07:27 INFO hive.HiveImport: Time taken:
    0.768 seconds
20/06/11 16:07:27 INFO hive.HiveImport:Hive import complete.
20/06/11 16:07:27 INFO hive.HiveImport:Export directory is con-
    tains the _SUCCESS file only,removing the directory.
```

执行完成之后，在 HDFS 中看到如图 6－11 所示的目录文件。

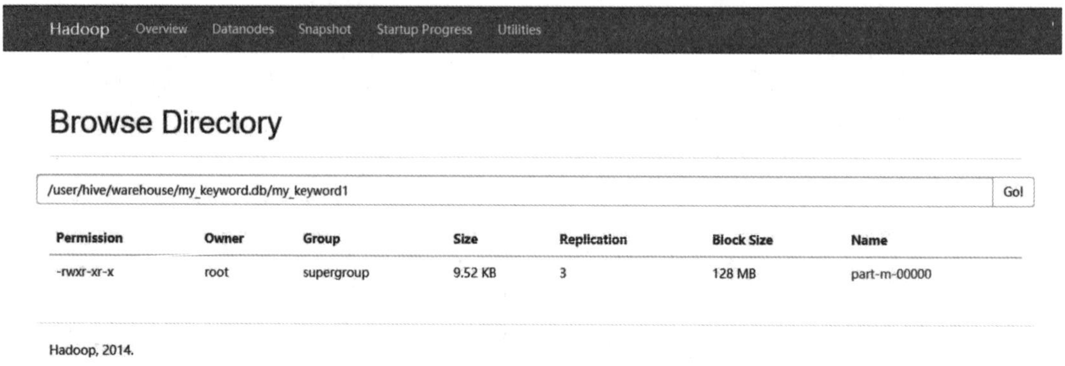

Hadoop　Overview　Datanodes　Snapshot　Startup Progress　Utilities

Browse Directory

/user/hive/warehouse/my_keyword.db/my_keyword1　Go!

Permission	Owner	Group	Size	Replication	Block Size	Name
-rwxr-xr-x	root	supergroup	9.52 KB	3	128 MB	part-m-00000

Hadoop, 2014.

图 6－11　指定行分隔符和列分隔符操作完成

通过命令查看 HDFS 文件数据如下：

```
[root@Superintendent2 ~]# hdfs dfs -cat /user/hive/warehouse/
  my_keyword.db/my_keyword1/part-m-00000
0,(JSON)
1, ->
2, ->>
3, < >
4,ACCOUNT
5,ACTION
6,ADD
7,AES_DECRYPT
8,AES_ENCRYPT
9,AFTER
 ... (中间部分省略)
698,ZEROFILL
```

通过 Hive 命令查看 Hive 数据表数据如下：

```
[root@Superintendent2 ~]# hive

```

```
Logging initialized using configuration in jar:file:/usr/local/hive/lib/hive-common-1.1.0.jar!/hive-log4j.properties

SLF4J:Class path contains multiple SLF4J bindings.
SLF4J:Found binding in[jar:file:/usr/local/hadoop/share/hadoop/common/lib/slf4j-log4j12-1.7.5.jar!/org/slf4j/impl/StaticLoggerBinder.class]
SLF4J:Found binding in[jar:file:/usr/local/hive/lib/hive-jdbc-1.1.0-standalone.jar!/org/slf4j/impl/StaticLoggerBinder.class]
SLF4J:See http://www.slf4j.org/codes.html#multiple_bindings for an explanation.
SLF4J:Actual binding is of type[org.slf4j.impl.Log4jLogger Factory]
hive>use my_keyword;
OK
Time taken:0.454 seconds
hive>show tables;
OK
my_keyword
Time taken:0.15 seconds,Fetched:1 row(s)
hive>select * from my_keyword;
OK
0	(JSON)
1 ->
2 ->>
3	< >
4	ACCOUNT
5	ACTION
6	ADD
7	AES_DECRYPT
8	AES_ENCRYPT
9	AFTER
...	(中间部分省略)
698	ZEROFILL
Time taken:0.354 seconds,Fetched:699 row(s)
hive>
```

3. Sqoop 数据导出

Sqoop 的 export 工具可以从 HDFS 同步一系列文件数据到 RDBMS 中。使用这个工具的前提是要导出的目标表在数据库中必须存在。导出文件根据用户指定的分隔符转化成一系列的输出记录。

默认的导出操作会将这些记录转化成一系列的 INSERT 语句，根据这些语句将记录插入关系型数据库中。而在 update 模式下，Sqoop 会生成一系列的 UPDATE 语句，将数据库中已经存在的记录进行更新。在 call 模式下，Sqoop 会为每一条记录调用一个存储过程来处理。Sqoop 的工作过程如图 6－12 所示。

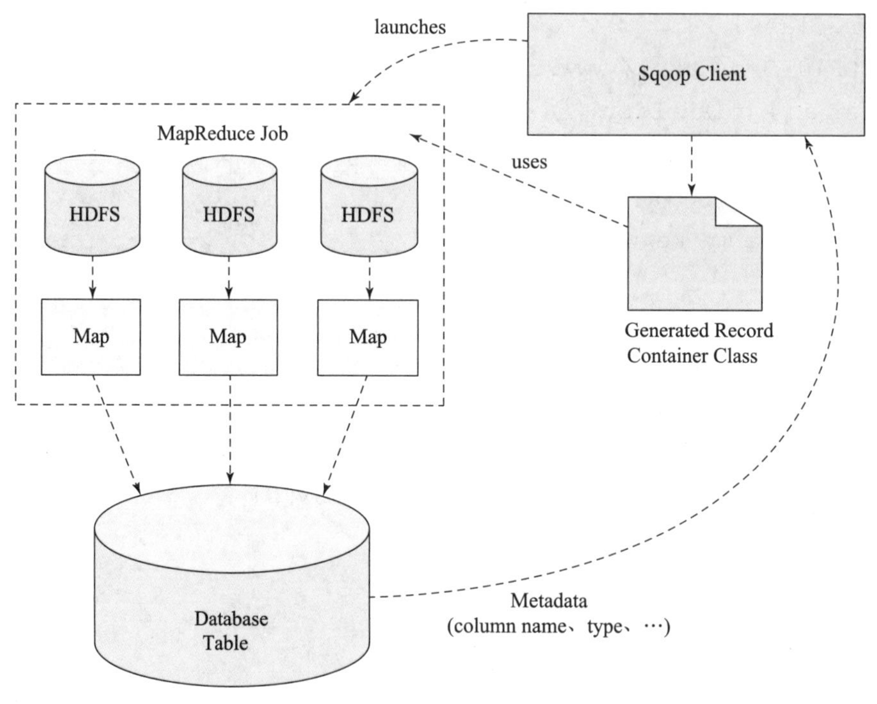

图 6－12　Sqoop 的工作过程

Sqoop 的 export 工具使用语法如下：

```
1.sqoop export[GENERIC-ARGS][TOOL-ARGS]
```

常见参数见表 6－5。

表 6－5　Sqoop 的常见参数

参数	描述
--connect <jdbc-uri>	JDBC 连接地址
--connection-manager <class-name>	连接管理者
--driver <class-name>	驱动类

续表

参数	描述
-- help	帮助信息
-- password -- file	为包含身份验证密码的文件设置路径
- P	从命令行输入密码
-- password < password >	密码
-- username < username >	用户名
-- verbose	打印流程信息
-- connection - param - file < filename >	可选参数
-- relaxed - isolation	将连接事务隔离设置为读取未提交的映射程序

验证参数见表 6 - 6。

表 6 - 6　验证参数

参数	描述
-- validate	启用验证的数据复制，只支持单表复制
-- validator < class - name >	指定要使用的 validator 类
-- validation - threshold < class - name >	指定验证入口所使用的类
-- validation - failurehandler < class - name >	指定要使用的验证失败处理类

控制参数见表 6 - 7。

表 6 - 7　控制参数

参数	描述
-- columns < col，col，col，… >	导出列表
-- direct	快速模式，利用了数据库的导入工具，如 MySQL 的 mysqlimport，可以比 JDBC 连接的方式更为高效地将数据导入到关系数据库中
-- export - dir < dir >	存放数据的 HDFS 的源目录
- m，-- num - mappers < n >	启动 n 个 Map 来并行导入数据，默认是 4 个，最好不要将数字设置为高于集群的最大 Map 数
-- table < table - name >	要导入到的关系数据库表
-- call < stored - proc - name >	要调用的存储过程
-- update - key < col - name >	后面接条件列名，通过该参数，可以将关系数据库中已经存在的数据进行更新操作，类似于关系数据库中的 update 操作
-- update - mode < mode >	更新模式，有两个值：updateonly 和默认的 allowinsert。该参数只有当关系数据表中不存在要导入的记录时才能使用，比如要导入的 HDFS 中有一条 id = 1 的记录，如果在表里已经有一条记录 id = 2，那么更新就会失败

续表

参数	描述
-- input - null - string <null - string>	可选参数，如果没有指定，则字符串 null 将被使用
-- input - null - none - string <null - string>	可选参数，如果没有指定，则字符串 null 将被使用
-- staging - table <staging - table - name>	该参数用来保证在数据导入关系数据库表的过程中事务的安全性。由于在导入的过程中可能会有多个事务，如果一个事务失败，则会影响到其他事务，比如导入的数据出现错误或出现重复的记录等，通过该参数可以避免这些情况。创建一个与导入目标表同样的数据结构，保留该表为空，在运行数据导入前，所有事务会将结果先存放在该表中，然后由该表通过一次事务将结果写入目标表中
-- clear - staging - table	如果该 staging - table 非空，则通过该参数可以在运行导入前清除 staging - table 里的数据
-- batch	使用批处理模式执行底层语句

-- export - dir、 -- table 和 -- call 是必须指定的参数。因为这三个参数用来指定导出数据集及数据导出后的去处，既可以是导出到某张表，也可以是对每一条导出记录调用存储过程。

默认情况下，会导出所有字段。可以使用 -- columns 参数指定部分字段进行导出。多个字段使用逗号分隔，比如 -- columns "col1,col2,col3"。需要注意的是，对那些没有指定的导出字段，在数据库中要么有默认值，要么运行为 NULL；否则，导出过程在插入这些字段的值时会报错。

在导出时，可以指定并行的 Map 个数，导出过程的速度由并发数来决定。默认情况下，export 过程会使用 4 个 Map 并行。具体的并发数由实际的数据量大小、数据库的性能等共同确定，由参数 -- num - mappers 或者 - m 来设置。

有些数据库提供一种 direct 模式，比如 MySQL。使用 -- direct 参数可以指定特定的数据库连接方式。

对于数据库中的 NULL 值定义，可以由参数 -- input - null - string 和 -- input - null - none - string 两个参数来指定。如果不指定 -- input - null - string 字段，那么导出数据中的 null 字符串会被当作 NULL 值插入数据库中；如果不指定 -- input - null - none - string，那么导出数据中的 null 字符串及空字符串都会被当作 NULL 值插入数据库中。这里需要注意对非 string 类型字段的操作提醒，会将空字符串直接导出为 NULL 值。

由于 Sqoop 将导出过程切分成多个 transaction，那么就有可能出现某个导出 job 失败而导致只有部分数据提交到了导出数据库中的情况。当失败 job 重试时，就有可能会出现数据重复或者导出数据冲突等情况，这时可以指定一个 -- staging - table 参数来避免这种情况的发生。导出的数据首先会缓存在该表中，等 job 执行成功后，会将该表中的数据移动到最终目标表中。

使用--staging-table时，需要在job执行前创建该表，这张表在结构上需要与目标表保持一致，并且这张表在任务执行前需要为空，或者指定--clear-staging-table参数。如果在导出前该表中有数据，那么这个参数会提前将该表清空。

--staging-table参数不能在--direct模式下工作，并且如果指定了--update-key或者指定的存储过程涉及数据的插入，则也不能使用--staging-table参数。

(1) 插入和更新

默认情况下，export操作会将记录插入导出表中。每一条导出记录都会转化成一条INSERT语句。如果导出表有字段约束（比如主键约束），则需要注意避免导出记录违反这些约束条件的限制。这种模式只适合将数据导出到一张新表或者空表中。

如果指定--update-key参数，Sqoop会将每一条导出记录转化成一条UPDATE语句，这样可以更新已存在的记录。比如，表定义语句如下：

```
CREATE TABLE foo(
    id INT NOT NULL PRIMARY KEY,
    msg VARCHAR(32),
    bar INT);
```

如果有一个HDFS文件内容如下：

```
0,this is a test,42
1,some more data,100
...
```

那么，运行sqoop-export --table foo --update-key id --export-dir/path/to/data --connect …，会执行一个export任务，这个任务执行的SQL语句如下：

```
UPDATE foo SET msg='this is a test',bar=42 WHERE id=0;
UPDATE foo SET msg='some more data',bar=100 WHERE id=1;
...
```

如果一条UPDATE语句匹配不到对应的记录或者匹配到多条记录，则不会发生任何错误，导出过程会正常继续执行后面的SQL语句。不过，如果在使用--update-key参数的同时指定参数--update-mode为allowinsert，则UPDATE语句匹配不到的记录就会以INSERT的形式插入目标表中。

--update-key参数也可以设置多个字段，多个字段之间用逗号隔开。

输入解析参数，见表6-8。

表6-8　解析参数

参数	描述
--input-enclosed-by <char>	设置一个必用的字段闭合符
--input-escaped-by <char>	设置转义符

续表

参数	描述
-- input - fields - terminated - by char >	设置字段分隔符
-- input - lines - terminated - by < char >	设置行结束符
-- input - optionally - enclosed - by < char >	设置一个可选的闭合符

输出行格式参数，见表6－9。

表6－9　行格式参数

参数	说明
-- enclosed - by < char >	设置一个必用的字段闭合符
-- escaped - by < char >	设置转义符
-- fields - terminated - by < char >	设置字段分隔符
-- lines - terminated - by < char >	设置行结束符
-- mysql - delimiters	使用 MySQL 默认的一组分隔符设置
-- optionally - enclosed - by < char >	设置一个字段闭合符（该闭合符只有在字段内出现分隔符时才会用于字段）

Sqoop 会根据上面设置的各种分隔符自动生成代码来解析输入文件。在非默认的情况下，需要设置这些参数来帮助 Sqoop 正确地解析输入文件中的记录。如果分隔符指定错误，则可能会出现对一行记录解析不到足够的字段数，从而报出 ParseExceptions 异常的情况。

代码生成参数，见表6－10。

表6－10　代码生成参数

参数	描述
-- bindir < dir >	指定生成的 Java 文件、编译成的 class 文件及将生成文件打包为 jar 的 jar 包文件输出路径
-- class - name < name >	设定生成的 Java 文件指定的名称
-- jar - file < file >	禁用代码生成；使用指定的 jar
-- outdir < dir >	生成的 Java 文件存放路径
-- package - name < name >	包名，如 cn. cnnic，则会生成 cn 和 cnnic 两级目录，生成的文件（如 Java 文件）存放在 cnnic 目录里
-- map - column - java < m >	数据库字段在生成的 Java 文件中会映射为各种属性，并且默认的数据类型与数据库类型保持对应，比如数据库中某字段的类型为 bigint，则在 Java 文件中的数据类型为 long。通过这个属性，可以改变数据库字段在 Java 中映射的数据类型，格式如 - map - column - java DB_ID = String, id = Integer

如果需要导入的数据已经由 import 导入，那么可以使用在导入时生成的代码来进行输

出，这时可以通过 --jar-file 和 --class-name 来指定使用的 jar 文件及需要使用的类名。

使用已有的代码时，会与 --update-key 参数相冲突，update 模式需要生成新的代码来完成这一过程。如果使用 update 模式，就不能使用 --jar-file 参数，并且必须指定所有没有默认值的分隔符。

(2) 导出和事务

数据的导出过程由多个 writer 并行执行，每一个 writer 使用一个独立的数据库连接，这些 writer 的 transaction 也是各自独立的。Sqoop 使用多个 INSERT 语句来插入记录，每一个 statement 最多能插入 100 条记录。每个 writer 在第 100 个 statement 时执行一次提交操作，所以，每一次提交可以插入 10 000 条记录。在导出过程结束前可以看到部分导出数据。

(3) 导出失败

数据可能会由于以下原因导致导出失败：

- Hadoop 集群失去与数据库的连接（可能是硬件原因或是网络等原因造成的）。
- 在 insert 模式下向有约束的表插入数据。
- HDFS 文件中记录本身有异常，数据无法正常解析。
- 分隔符指定错误，导致解析发生错误。
- 硬件原因，比如磁盘空间不足或内存不足等。

如果一个导出 export 的数据的 Map 任务失败，将会导致整个 export 任务失败，并且失败的 export 过程的结果是未知的。每个 Map 任务在独立的 transaction 中并发地导出数据，并且每一个 Map 任务定期提交当前 transaction 中的记录。如果一个 Map 任务失败，当前 transaction 将会回滚，而之前正常提交的 transaction 中的记录已经导出到外部表中了。

下面通过几个实际操作来巩固知识点。

实例 1：将 HDFS 上的/sqoop/YYR/myoutport1 数据导出到 MySQL 中。

首先，在 MySQL 中创建表：

```
# 切换数据库
mysql >use mysql;
Reading table information for completion of table and column names
You can turn off this feature to get a quicker startup with -A
Database changed
# 查看 help_keyword 表结构
mysql >desc help_keyword;
+--------------+---------------+----+-----+-----+----+
|Field         |Type           |Null|Key  |Default|Extra|
+--------------+---------------+----+-----+-----+----+
```

```
|help_keyword_id |int(10) unsigned  |NO  |PRI  |NULL |     |
|name            |char(64)          |NO  |UNI  |NULL |     |
+----------------+------------------+----+-----+-----+-----+
2 rows in set (0.00 sec)

#创建数据库,用来导出数据
mysql >create database my_keyword;
Query OK,1 row affected (0.00 sec)

#切换为新创建的数据
mysql >use my_keyword;
Database changed

#创建导出数据目标表
mysql >create table my_keyword(id int(10),name char(64));
Query OK,0 rows affected (0.19 sec)

#确认创建成功
mysql >show tables;
+--------------------+
|Tables_in_my_keyword |
+--------------------+
|my_keyword          |
+--------------------+
1 row in set (0.00 sec)

#查看表结构
mysql >desc my_keyword;
+------+---------+-----+----+--------+------+
|Field |Type     |Null |Key |Default |Extra |
+------+---------+-----+----+--------+------+
|id    |int(10)  |YES  |    |NULL    |      |
|name  |char(64) |YES  |    |NULL    |      |
+------+---------+-----+----+--------+------+
2 rows in set (0.00 sec)
```

通过 Sqoop 将数据导出到 MySQL 数据表:

```
# 数据导出命令
[root@Superintendent2 ~]# sqoop export --connect jdbc:mysql://YYR:3306/my_keyword --username root --password root00 -m 1 --columns id,name --export-dir /user/YYR/my_help_keyword --table my_keyword
# 数据导出命令执行过程如下
20/06/11 16:24:13 INFO sqoop.Sqoop:Running Sqoop version:1.4.7
20/06/11 16:24:13 WARN tool.BaseSqoopTool:Setting your password on the command-line is insecure. Consider using -P instead.
20/06/11 16:24:13 INFO manager.MySQLManager:Preparing to use a MySQL streaming resultset.
20/06/11 16:24:13 INFO tool.CodeGenTool:Beginning code generation
Loading class 'com.mysql.jdbc.Driver'. This is deprecated. The new driver class is 'com.mysql.cj.jdbc.Driver'. The driver is automatically registered via the SPI and manual loading of the driver class is generally unnecessary.
20/06/11 16:24:14 INFO manager.SqlManager:Executing SQL statement:SELECT t.* FROM 'my_keyword' AS t LIMIT 1
20/06/11 16:24:14 INFO manager.SqlManager:Executing SQL statement:SELECT t.* FROM 'my_keyword' AS t LIMIT 1
20/06/11 16:24:14 INFO orm.CompilationManager:HADOOP_MAPRED_HOME is /usr/local/hadoop
注:/tmp/sqoop-root/compile/3766b9ddda920dbd8c1ec3e91a6ce4cb/my_keyword.java 使用或覆盖了已过时的 API。
注:有关的详细信息,请使用 -Xlint:deprecation 重新编译。
20/06/11 16:24:15 INFO orm.CompilationManager:Writing jar file:/tmp/sqoop-root/compile/3766b9ddda920dbd8c1ec3e91a6ce4cb/my_keyword.jar
20/06/11 16:24:15 INFO mapreduce.ExportJobBase:Beginning export of my_keyword
20/06/11 16:24:15 INFO Configuration.deprecation:mapred.jar is deprecated. Instead,use mapreduce.job.jar
20/06/11 16:24:16 INFO Configuration.deprecation:mapred.reduce.tasks.speculative.execution is deprecated. Instead, use mapreduce.reduce.speculative
```

```
20/06/11 16:24:16 INFO Configuration.deprecation:mapred.map.tasks.speculative.execution is deprecated. Instead,use mapreduce.map.speculative
20/06/11 16:24:16 INFO Configuration.deprecation:mapred.map.tasks is deprecated. Instead,use mapreduce.job.maps
20/06/11 16:24:16 INFO client.RMProxy:Connecting to ResourceManager at YYR/192.168.3.190:8032
20/06/11 16:24:18 INFO input.FileInputFormat:Total input paths to process:4
20/06/11 16:24:18 INFO input.FileInputFormat:Total input paths to process:4
20/06/11 16:24:18 INFO mapreduce.JobSubmitter: number of splits:1
20/06/11 16:24:18 INFO Configuration.deprecation:mapred.map.tasks.speculative.execution is deprecated. Instead,use mapreduce.map.speculative
20/06/11 16:24:18 INFO mapreduce.JobSubmitter:Submitting tokens for job:job_1557902299589_0020
20/06/11 16:24:18 INFO impl.YarnClientImpl:Submitted application application_1557902299589_0020
20/06/11 16:24:18 INFO mapreduce.Job:The url to track the job:http://YYR:8088/proxy/application_1557902299589_0020/
20/06/11 16:24:18 INFO mapreduce.Job:Running job:job_1557902299589_0020
20/06/11 16:24:24 INFO mapreduce.Job:Job job_1557902299589_0020 running in uber mode:false
20/06/11 16:24:24 INFO mapreduce.Job:map 0%  reduce 0%
20/06/11 16:24:29 INFO mapreduce.Job:map 100%  reduce 0%
20/06/11 16:24:29 INFO mapreduce.Job:Job job_1557902299589_0020 completed successfully
20/06/11 16:24:29 INFO mapreduce.Job:Counters:30
        File System Counters
                FILE:Number of bytes read=0
                FILE:Number of bytes written=123805
                FILE:Number of read operations=0
                FILE:Number of large read operations=0
                FILE:Number of write operations=0
```

```
                HDFS:Number of bytes read =10128
                HDFS:Number of bytes written =0
                HDFS:Number of read operations =13
                HDFS:Number of large read operations =0
                HDFS:Number of write operations =0
        Job Counters
                Launched map tasks =1
                Rack -local map tasks =1
                Total time spent by all maps in occupied slots (ms) =3010
                Total time spent by all reduces in occupied slots (ms) =0
                Total time spent by all map tasks (ms) =3010
                Total vcore -seconds taken by all map tasks =3010
                Total megabyte -seconds taken by all map tasks =3082240
        Map -Reduce Framework
                Map input records =699
                Map output records =699
                Input split bytes =368
                Spilled Records =0
                Failed Shuffles =0
                Merged Map outputs =0
                GC time elapsed (ms) =50
                CPU time spent (ms) =1670
                Physical memory (bytes) snapshot =178081792
                Virtual memory (bytes) snapshot =2138980352
                Total committed heap usage (bytes) =147849216
        File Input Format Counters
                Bytes Read =0
        File Output Format Counters
                Bytes Written =0
20/06/11 16:24:29 INFO mapreduce.ExportJobBase: Transferred 9.8906 KB in 13.7459 seconds (736.8033 bytes/sec)
20/06/11 16:24:29 INFO mapreduce.ExportJobBase:Exported 699 records.
```

查看 MySQL 数据表中的数据:

```
mysql >select * from my_keyword;
+------+---------------------------------+
| id   | name                            |
+------+---------------------------------+
|    0 | (JSON)                          |
|    1 | ->                              |
|    2 | ->>                             |
|    3 | < >                             |
|    4 | ACCOUNT                         |
|    5 | ACTION                          |
|    6 | ADD                             |
|    7 | AES_DECRYPT                     |
|    8 | AES_ENCRYPT                     |
|    9 | AFTER                           |
| ...  (中间部分省略)                    |
|  698 | ZEROFILL                        |
+------+---------------------------------+
699 rows in set (0.00 sec)

mysql >
```

实例2：将Hive中的数据导出到MySQL中。

首先，在MySQL中创建表：

```
mysql >create table my_keyword1(id int(10),name char(64));
Query OK,0 rows affected (0.22 sec)
```

通过Sqoop将Hive数据导出到MySQL数据表：

```
# 数据导出命令
[root@Superintendent2 ~]# sqoop export --connect jdbc:mysql://YYR:3306/my_keyword --username root --password root00 -m 1 --columns id,name --export-dir/user/hive/warehouse/my_keyword.db/my_keyword1 --table my_keyword1 --input-fields-terminated-by ","
# 数据导出命令执行过程如下
20/06/11 16:30:07 INFO sqoop.Sqoop:Running Sqoop version:1.4.7
20/06/11 16:30:07 WARN tool.BaseSqoopTool:Setting your password on the command-line is insecure. Consider using -P instead.
```

```
6. 20/06/11 16:30:07 INFO manager.MySQLManager:Preparing to use a
   MySQL streaming resultset.
7.20/06/11 16:30:07 INFO tool.CodeGenTool:Beginning code genera-
   tion
8.Loading class 'com.mysql.jdbc.Driver'. This is deprecated. The
   new driver class is 'com.mysql.cj.jdbc.Driver'. The driver is au-
   tomatically registered via the SPI and manual loading of the
   driver class is generally unnecessary.
9.20/06/11 16:30:07 INFO manager.SqlManager:Executing SQL state-
   ment:SELECT t.* FROM 'my_keyword1' AS t LIMIT 1
10. 20/06/11 16:30:07 INFO manager.SqlManager:Executing SQL state-
    ment:SELECT t.* FROM 'my_keyword1' AS t LIMIT 1
11.20/06/11 16:30:07 INFO orm.CompilationManager:HADOOP_MAPRED_
    HOME is/usr/local/hadoop
12. 注:/tmp/sqoop-root/compile/dd216aad0edd0e39b12d970fea5858ac/
    my_keyword1.java 使用或覆盖了已过时的 API。
13. 注:有关的详细信息,请使用 -Xlint:deprecation 重新编译。
14. 20/06/11 16:30:09 INFO orm.CompilationManager:Writing jar file:/
    tmp/sqoop-root/compile/dd216aad0edd0e39b12d970fea5858ac/my_
    keyword1.jar
15.20/06/11 16:30:09 INFO mapreduce.ExportJobBase:Beginning ex-
    port of my_keyword1
16.20/06/11 16:30:09 INFO Configuration.deprecation:mapred.jar is
    deprecated. Instead,use mapreduce.job.jar
17.20/06/11 16:30:10 INFO Configuration.deprecation:mapred. re-
    duce.tasks.speculative.execution is deprecated. Instead, use
    mapreduce.reduce.speculative
18.20/06/11 16:30:10 INFO Configuration.deprecation: mapred.
    map.tasks.speculative.execution is deprecated. Instead, use
    mapreduce.map.speculative
19.20/06/11 16:30:10 INFO Configuration.deprecation: mapred.
    map.tasks is deprecated. Instead,use mapreduce.job.maps
20.20/06/11 16:30:10 INFO client.RMProxy:Connecting to Resource
    Manager at YYR/192.168.3.190:8032
21.20/06/11 16:30:11 INFO input.FileInputFormat:Total input paths
    to process:1
```

```
20/06/11 16:30:11 INFO input.FileInputFormat:Total input paths to process:1
20/06/11 16:30:11 INFO mapreduce.JobSubmitter: number of splits:1
20/06/11 16:30:11 INFO Configuration.deprecation:mapred.map.tasks.speculative.execution is deprecated. Instead,use mapreduce.map.speculative
20/06/11 16:30:12 INFO mapreduce.JobSubmitter:Submitting tokens for job:job_1557902299589_0021
20/06/11 16:30:12 INFO impl.YarnClientImpl:Submitted application application_1557902299589_0021
20/06/11 16:30:12 INFO mapreduce.Job:The url to track the job:http://YYR:8088/proxy/application_1557902299589_0021/
20/06/11 16:30:12 INFO mapreduce.Job: Running job: job_1557902299589_0021
20/06/11 16:30:18 INFO mapreduce.Job:Job job_1557902299589_0021 running in uber mode:false
20/06/11 16:30:18 INFO mapreduce.Job:map 0% reduce 0%
20/06/11 16:30:24 INFO mapreduce.Job:map 100% reduce 0%
20/06/11 16:30:24 INFO mapreduce.Job:Job job_1557902299589_0021 completed successfully
20/06/11 16:30:24 INFO mapreduce.Job:Counters:30
        File System Counters
                FILE:Number of bytes read=0
                FILE:Number of bytes written=123826
                FILE:Number of read operations=0
                FILE:Number of large read operations=0
                FILE:Number of write operations=0
                HDFS:Number of bytes read=9909
                HDFS:Number of bytes written=0
                HDFS:Number of read operations=4
                HDFS:Number of large read operations=0
                HDFS:Number of write operations=0
        Job Counters
                Launched map tasks=1
                Rack-local map tasks=1
```

```
                Total time spent by all maps in occupied slots (ms)=2968
                Total time spent by all reduces in occupied slots (ms)=0
                Total time spent by all map tasks (ms)=2968
                Total vcore-seconds taken by all map tasks=2968
                Total megabyte-seconds taken by all map tasks=3039232
        Map-Reduce Framework
                Map input records=699
                Map output records=699
                Input split bytes=158
                Spilled Records=0
                Failed Shuffles=0
                Merged Map outputs=0
                GC time elapsed (ms)=48
                CPU time spent (ms)=1570
                Physical memory (bytes) snapshot=175644672
                Virtual memory (bytes) snapshot=2136281088
                Total committed heap usage (bytes)=143654912
        File Input Format Counters
                Bytes Read=0
        File Output Format Counters
                Bytes Written=0
20/06/11 16:30:24 INFO mapreduce.ExportJobBase: Transferred 9.6768 KB in 14.5668 seconds (680.2458 bytes/sec)
20/06/11 16:30:24 INFO mapreduce.ExportJobBase: Exported 699 records.
```

查看 MySQL 数据表中的数据：

```
mysql>select * from my_keyword1;
+------+----------------------------------+
|id    |name                              |
+------+----------------------------------+
|    0 |(JSON)                            |
|    1 |->                                |
|    2 |->>                               |
```

```
|     3 |< >                              |
|     4 |ACCOUNT                           |
|     5 |ACTION                            |
|     6 |ADD                               |
|     7 |AES_DECRYPT                       |
|     8 |AES_ENCRYPT                       |
|     9 |AFTER                             |
|   ...  (中间部分省略)                    |
|   698 |ZEROFILL                          |
+-----+--------------------------------+
699 rows in set (0.00 sec)

mysql >
```

运行上述程序，Sqoop 组件运行正常。

学习笔记

项目七
Azkaban 调度器

项目描述

模拟大数据业务处理场景时，通常会遇到这样的场景：

A 任务：将收集的数据通过一系列的规则进行清洗，然后存入 Hive 的表 a 中。

B 任务：将 Hive 中已存在的表 b 和表 c 进行关联，得到表 d。

C 任务：将 A 任务中得到的表 a 与 B 任务中得到的表 d 进行关联，得到分析的结果表 e。

D 任务：将 Hive 中得到的表 e 通过 Sqoop 导入关系型数据库 MySQL 中，以供 Web 端查询使用。

显然，任务 C 依赖于任务 A 与任务 B 的结果，任务 D 依赖于任务 C 的结果。通常的做法是打开两个终端，分别执行任务 A 与任务 B，当任务 A 与任务 B 执行完成之后，再执行任务 C，当任务 C 执行完成之后，再执行任务 D。整个任务流程中，必须保证任务 A、任务 B 执行完成之后再执行任务 C，然后再执行任务 D。这个操作过程中，每个环节都需要人工参与，时刻盯着各任务的执行进度。

项目分析

本项目主要通过以下几个任务来完成调度器 Azkaban 的学习：

任务 1：Azkaban 的安装部署。

任务 2：导入数据库。

任务 3：验证 Azkaban。

任务 4：构建工作流。

这个业务场景就是一个相对较大的任务，任务中分为四个子任务 A、B、C、D，如果能够通过任务调度器自动按照要求完成各个任务，就能大大降低人员参与度，从而降低系统维护成本。Azkaban 就是这样一个工作流的调度器，可以解决以上场景问题。

任务 1　Azkaban 的安装部署

一个完整的数据分析系统通常由多个前后依赖的模块组合构成：数据采集、数据预处理、

数据分析、数据展示等。各个模块单元之间存在时间先后依赖关系，并且存在着周期性重复。

为了很好地执行这样复杂的计划，需要一个工作流调度系统来调度执行，这里选择使用 Azkaban 调度器。

1. 安装 Azkaban

Azkaban 是一个批量工作流调度器，底层是使用 Java 语言开发的，用于在一个工作流内以一定的顺序运行一组任务和流程，并且提供了非常方便的 Web UI 界面来监控任务调度的情况，方便管理流调度任务。

组成 Azkaban 的三个关键组件如下：

- AzkabanWebServer：

  主要负责项目管理、用户登录权限认证、定时执行工作任务、跟踪和提交任务执行的流程、访问历史执行任务、保存执行计划的状态。
- AzkabanExecutorServer

  主要负责工作流程的提交、执行、检索，更新当前正在执行计划的数据，以及处理执行计划的日志。
- 关系型数据库

  主要是保存工作流中的原数据信息。

下面开始搭建一个 Azkaban 工作流调度系统。

(1) 下载 Azkaban

下载 AZKABAN

登录 Azkaban 的官网 https://azkaban.github.io/，如图 7－1 所示，单击“Downloads”按钮。

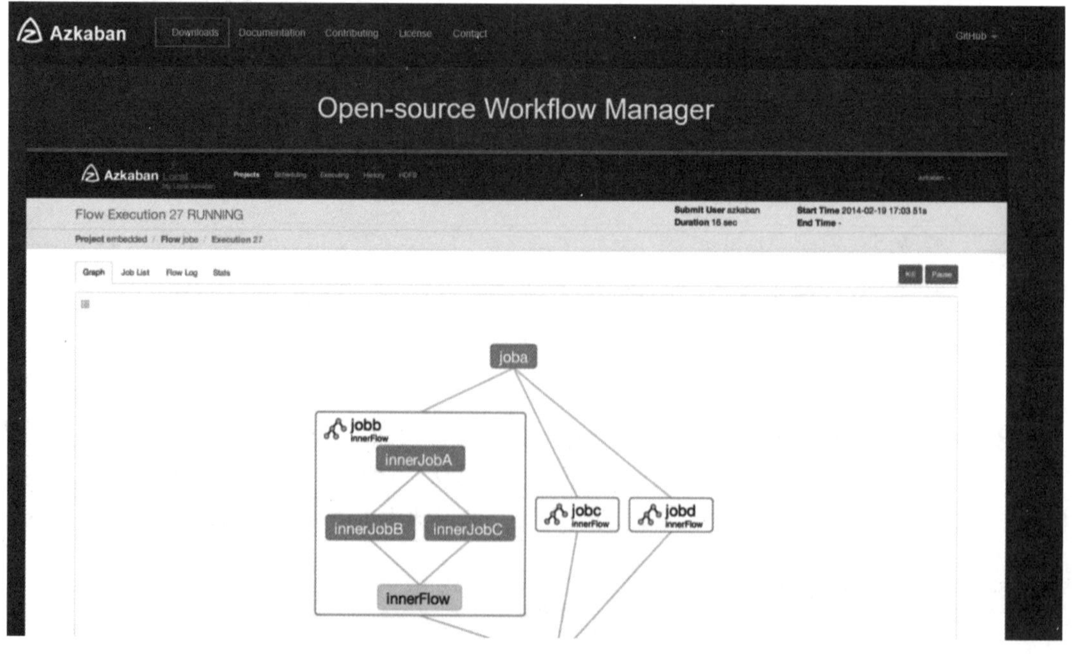

图 7－1 官网下载 Azkaban

在跳转的页面中选择“Releases”，进入页面并选择相应的版本进行下载，这里选择的是3.70.0版本，单击“Source code(tar.gz)”进行下载，如图7-2和图7-3所示。

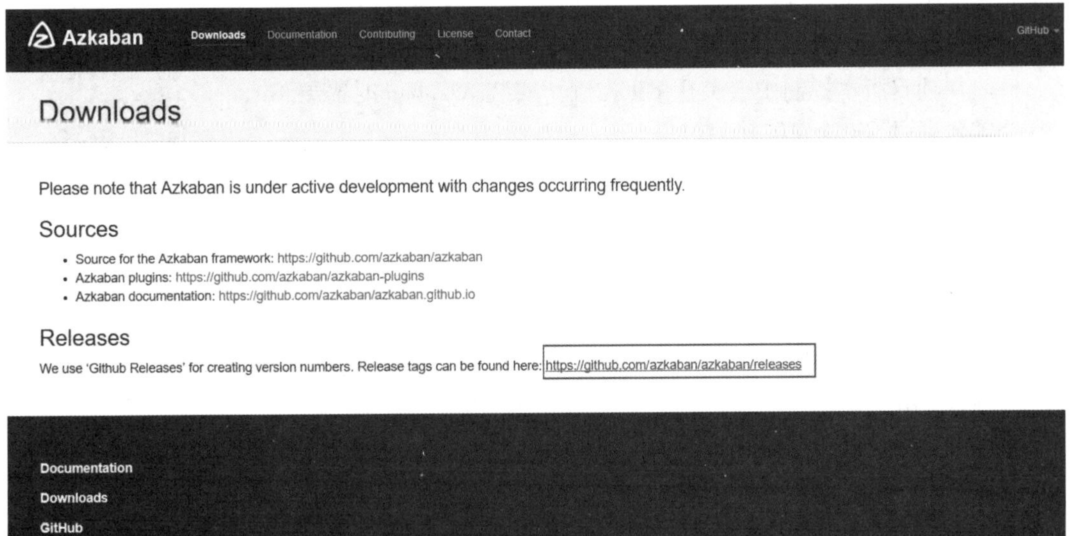

图7-2　选择“Releases”

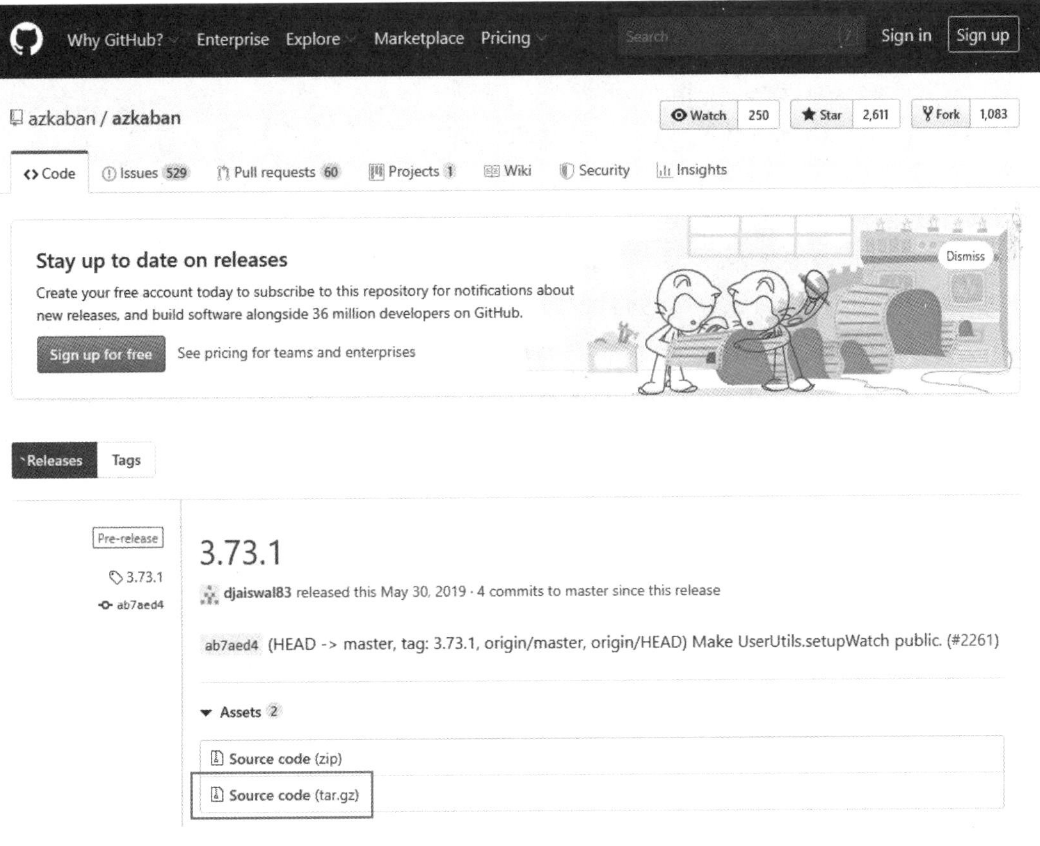

图7-3　版本选择

（2）环境准备

要在 Linux 中安装 Azkaban，需要在系统中准备好 JDK、MySQL，这里选择的是 JDK 8 和 MySQL 5.7 版本。由于已在 YYR 主机完成了部署，这里就不再重复安装过程了。另外，还需要安装 Git。Git 是一个开源的分布式版本控制系统，一般在项目版本控制中会使用 Git 控制。安装 Azkaban 时需要 Git，是因为需要通过 Git 构建依赖的包。

环境准备

安装 Git，执行命令如下：

```
# 安装 Git
[root@YYR ~]# yum install git
已加载插件:fastestmirror
Base                                              |3.6 kB  00:00:00
extras                                            |3.4 kB  00:00:00
mysql-connectors-community                        |2.5 kB  00:00:00
mysql-tools-community                             |2.5 kB  00:00:00
mysql57-community                                 |2.5 kB  00:00:00
updates                                           |3.4 kB  00:00:00
(1/2):extras/7/x86_64/primary_db                  |204 kB  00:00:00
(2/2):updates/7/x86_64/primary_db                 |6.4 MB  00:00:02
Determining fastest mirrors
 * base:mirrors.huaweicloud.com
 * extras:mirrors.nwsuaf.edu.cn
 * updates:mirrors.nwsuaf.edu.cn
正在解决依赖关系
--> 正在检查事务
---> 软件包 git.x86_64.0.1.8.3.1-20.el7 将被安装
--> 正在处理依赖关系 perl-Git = 1.8.3.1-20.el7,它被软件包 git-1.8.3.1-20.el7.x86_64 需要
--> 正在处理依赖关系 rsync,它被软件包 git-1.8.3.1-20.el7.x86_64 需要
--> 正在处理依赖关系 perl(Term::ReadKey),它被软件包 git-1.8.3.1-20.el7.x86_64 需要
--> 正在处理依赖关系 perl(Git),它被软件包 git-1.8.3.1-20.el7.x86_64 需要
--> 正在处理依赖关系 perl(Error),它被软件包 git-1.8.3.1-20.el7.x86_64 需要
--> 正在检查事务
---> 软件包 perl-Error.noarch.1.0.17020-2.el7 将被安装
---> 软件包 perl-Git.noarch.0.1.8.3.1-20.el7 将被安装
```

```
---> 软件包 perl-TermReadKey.x86_64.0.2.30-20.el7 将被安装
---> 软件包 rsync.x86_64.0.3.1.2-6.el7_6.1 将被安装
--> 解决依赖关系完成

依赖关系解决

==========================================================================
Package                 架构              版本                源        大小
==========================================================================
正在安装:
git                  x86_64   1.8.3.1-20.el7     updates    4.4 MB
为依赖而安装:
perl-Error           noarch   1:0.17020-2.el7    base        32 KB
perl-Git             noarch   1.8.3.1-20.el7     updates     55 KB
perl-TermReadKey     x86_64   2.30-20.el7        base        31 KB
rsync                x86_64   3.1.2-6.el7_6.1    updates    404 KB

事务概要
==========================================================================
安装  1 软件包(+4 依赖软件包)

总下载量:4.9 MB
安装大小:23 MB
Is this ok[y/d/N]:y
Downloading packages:
(1/5):perl-Error-0.17020-2.el7.noarch.rpm
                                                  |  32 KB  00:00:00
(2/5):perl-Git-1.8.3.1-20.el7.noarch.rpm          |  55 KB  00:00:00
(3/5):perl-TermReadKey-2.30-20.el7.x86_64.rpm
                                                  |  31 KB  00:00:00
(4/5):rsync-3.1.2-6.el7_6.1.x86_64.rpm            | 404 KB  00:00:01
(5/5):git-1.8.3.1-20.el7.x86_64.rpm               | 4.4 MB  00:00:02
-------------------------------------------------------------------------
总计                                     2.1 MB/s | 4.9 MB  00:00:02
Running transaction check
Running transaction test
```

```
Transaction test succeeded
Running transaction
  正在安装:1:perl-Error-0.17020-2.el7.noarch                    1/5
  正在安装:rsync-3.1.2-6.el7_6.1.x86_64                         2/5
  正在安装:perl-TermReadKey-2.30-20.el7.x86_64                  3/5
  正在安装:git-1.8.3.1-20.el7.x86_64                            4/5
  正在安装:perl-Git-1.8.3.1-20.el7.noarch                       5/5
  验证中:perl-Git-1.8.3.1-20.el7.noarch                         1/5
  验证中:1:perl-Error-0.17020-2.el7.noarch                      2/5
  验证中:perl-TermReadKey-2.30-20.el7.x86_64                    3/5
  验证中:git-1.8.3.1-20.el7.x86_64                              4/5
  验证中:rsync-3.1.2-6.el7_6.1.x86_64                           5/5

已安装:
  git.x86_64 0:1.8.3.1-20.el7

作为依赖被安装:
  perl-Error.noarch 1:0.17020-2.el7    perl-Git.noarch 0:1.8.3.1-20.el7    perl-TermReadKey.x86_64 0:2.30-20.el7
    rsync.x86_64 0:3.1.2-6.el7_6.1

完毕!

# 查看 Git 版本号
[root@YYR ~]# git --version
git version 1.8.3.1
```

(3) 安装 Azkaban

①上传下载好的 Azkaban，解压到/usr/local/文件夹。

```
# 解压缩
[root@YYR ~]# tar -zxvf /tmp/azkaban-3.73.1.tar.gz -C /usr/local/
azkaban-3.73.1/
azkaban-3.73.1/.gitignore
azkaban-3.73.1/.travis.yml
azkaban-3.73.1/CONTRIBUTING.md
azkaban-3.73.1/LICENSE
```

```
8.azkaban-3.73.1/NOTICE
9.azkaban-3.73.1/README.md
10.azkaban-3.73.1/az-core/
11.azkaban-3.73.1/az-core/build.gradle
12.azkaban-3.73.1/az-core/src/
13.azkaban-3.73.1/az-core/src/main/
14.azkaban-3.73.1/az-core/src/main/java/
15.azkaban-3.73.1/az-core/src/main/java/azkaban/
16. azkaban-3.73.1/az-core/src/main/java/azkaban/AzkabanCore-
   Module.java
17.azkaban-3.73.1/az-core/src/main/java/azkaban/Constants.
   java
18.azkaban-3.73.1/az-core/src/main/java/azkaban/metrics/
19.azkaban-3.73.1/az-core/src/main/java/azkaban/metrics/Met-
   ricsManager.java
20.azkaban-3.73.1/az-core/src/main/java/azkaban/utils/
21.azkaban-3.73.1/az-core/src/main/java/azkaban/utils/Execu-
   torServiceUtils.java
22.azkaban-3.73.1/az-core/src/main/java/azkaban/utils/JSONU-
   tils.java
23. azkaban-3.73.1/az-core/src/main/java/azkaban/utils/Mem-
   ConfValue.java
24.azkaban-3.73.1/az-core/src/main/java/azkaban/utils/Plugi-
   nUtils.java
25.azkaban-3.73.1/az-core/src/main/java/azkaban/utils/Props.
   java
26.azkaban-3.73.1/az-core/src/main/java/azkaban/utils/Prop-
   sUtils. java
27.azkaban-3.73.1/az-core/src/main/java/azkaban/utils/TimeU-
   tils.java
28.azkaban-3.73.1/az-core/src/main/java/azkaban/utils/Unde-
   finedPropertyException.java
29.azkaban-3.73.1/az-core/src/main/java/azkaban/utils/
   Utils.java
30.azkaban-3.73.1/az-core/src/test/
31.azkaban-3.73.1/az-core/src/test/java/
32. azkaban-3.73.1/az-core/src/test/java/azkaban/
```

```
33. azkaban-3.73.1/az-core/src/test/java/azkaban/metrics/
34.... （中间过程省略）
35.azkaban-3.73.1/tools/
36.azkaban-3.73.1/tools/README.md
37.azkaban-3.73.1/tools/create_release.py
38.azkaban-3.73.1/tools/create_release_test.py
```

②修改文件名。

```
1. [root@YYR ~]# mv/usr/local/azkaban-3.73.1 //usr/local/azka-
   ban
```

③编译 Azkaban。

编译 AZKABAN

```
1.# 进入 Azkaban 目录
2.[root@YYR ~]# cd/usr/local/azkaban/
3.# 编译
4.[root@YYR azkaban]# ./gradlew disTar
5. Parallel execution with configuration on demand is an incubating
   feature.
6.Download https:// repo.maven.apache.org/ maven2/ log4j/ log4j/
   1.2.16/log4j-1.2.16.pom
7. Download https://repo.maven.apache.org/maven2/org/slf4j/slf4j
   -api/1.7.18/slf4j-api-1.7.18.pom
8.Download https:// repo.maven.apache.org/ maven2/ com/ linkedin/
   pegasus/generator/1.15.7/generator-1.15.7.pom
9.Download https://repo.maven.apache.org/maven2/org/slf4j/slf4j-
   parent/1.7.18/slf4j-parent-1.7.18.pom
10.Download https:// repo.maven.apache.org/ maven2/ com/ linkedin/
   pegasus/restli-tools/1.15.7/restli-tools-1.15.7.pom
11.Download https://repo.maven.apache.org/maven2/joda-time/jo-
   da-time/2.0/joda-time-2.0.pom
12.Download https://repo.maven.apache.org/maven2/commons-lang/
   commons-lang/2.6/commons-lang-2.6.pom
13.Download https://repo.maven.apache.org/maven2/com/google/in-
   ject/guice/4.1.0/guice-4.1.0.pom
14.Download https:// repo.maven.apache.org/ maven2/ org/ slf4j/
   slf4j-api/1.6.2/slf4j-api-1.6.2.pom
15.Download https:// repo.maven.apache.org/ maven2/ org/ apache/
   commons/commons-parent/17/commons-parent-17.pom
```

16. Download https://repo.maven.apache.org/maven2/com/google/inject/guice-parent/4.1.0/guice-parent-4.1.0.pom
17. Download https://repo.maven.apache.org/maven2/org/apache/commons/commons-jexl/2.1.1/commons-jexl-2.1.1.pom
18. Download https://repo.maven.apache.org/maven2/org/slf4j/slf4j-parent/1.6.2/slf4j-parent-1.6.2.pom
19. Download https://repo.maven.apache.org/maven2/com/google/google/5/google-5.pom
20. Download https://repo.maven.apache.org/maven2/org/codehaus/jackson/jackson-core-asl/1.9.5/jackson-core-asl-1.9.5.pom
21. Download https://repo.maven.apache.org/maven2/org/quartz-scheduler/quartz/2.2.1/quartz-2.2.1.pom
22. Download https://repo.maven.apache.org/maven2/com/linkedin/pegasus/pegasus-common/1.15.7/pegasus-common-1.15.7.pom
23. Download https://repo.maven.apache.org/maven2/io/dropwizard/metrics/metrics-jvm/3.1.0/metrics-jvm-3.1.0.pom
24. Download https://repo.maven.apache.org/maven2/javax/inject/javax.inject/1/javax.inject-1.pom
25. Download https://repo.maven.apache.org/maven2/org/apache/apache/7/apache-7.pom
26. Download https://repo.maven.apache.org/maven2/org/quartz-scheduler/quartz-parent/2.2.1/quartz-parent-2.2.1.pom
27. Download https://repo.maven.apache.org/maven2/io/dropwizard/metrics/metrics-parent/3.1.0/metrics-parent-3.1.0.pom
28. Download https://repo.maven.apache.org/maven2/aopalliance/aopalliance/1.0/aopalliance-1.0.pom
29. Download https://repo.maven.apache.org/maven2/org/slf4j/slf4j-log4j12/1.6.2/slf4j-log4j12-1.6.2.pom
30. Download https://repo.maven.apache.org/maven2/com/google/guava/guava/21.0/guava-21.0.pom
31. Download https://repo.maven.apache.org/maven2/com/google/guava/guava/19.0/guava-19.0.pom
32. Download https://repo.maven.apache.org/maven2/com/linkedin/pegasus/data/1.15.7/data-1.15.7.pom
33. Download https://repo.maven.apache.org/maven2/commons-cli/commons-cli/1.3.1/commons-cli-1.3.1.pom

```
34. Download https://repo.maven.apache.org/maven2/com/google/guava/guava-parent/21.0/guava-parent-21.0.pom
35. Download https://repo.maven.apache.org/maven2/io/dropwizard/metrics/metrics-core/3.1.0/metrics-core-3.1.0.pom
36. Download https://repo.maven.apache.org/maven2/com/fasterxml/jackson/core/jackson-core/2.9.2/jackson-core-2.9.2.pom
37. Download https://repo.maven.apache.org/maven2/com/fasterxml/jackson/jackson-base/2.9.2/jackson-base-2.9.2.pom
38. Download https://repo.maven.apache.org/maven2/commons-io/commons-io/2.4/commons-io-2.4.pom
39. Download https://repo.maven.apache.org/maven2/org/apache/commons/commons-parent/37/commons-parent-37.pom
40. Download https://repo.maven.apache.org/maven2/org/codehaus/jackson/jackson-mapper-asl/1.9.5/jackson-mapper-asl-1.9.5.pom
41. Download https://repo.maven.apache.org/maven2/com/fasterxml/jackson/jackson-bom/2.9.2/jackson-bom-2.9.2.pom
42. Download https://repo.maven.apache.org/maven2/org/apache/hadoop/hadoop-common/2.6.1/hadoop-common-2.6.1.pom
43. Download https://repo.maven.apache.org/maven2/com/linkedin/pegasus/r2/1.15.7/r2-1.15.7.pom
44. Download https://repo.maven.apache.org/maven2/org/bouncycastle/bcprov-jdk15on/1.54/bcprov-jdk15on-1.54.pom
45. Download https://repo.maven.apache.org/maven2/org/apache/commons/commons-parent/25/commons-parent-25.pom
46. Download https://repo.maven.apache.org/maven2/com/fasterxml/jackson/jackson-parent/2.9.1/jackson-parent-2.9.1.pom
47. Download https://repo.maven.apache.org/maven2/com/fasterxml/jackson/core/jackson-annotations/2.7.4/jackson-annotations-2.7.4.pom
48. Download https://repo.maven.apache.org/maven2/com/google/guava/guava-parent/19.0/guava-parent-19.0.pom
49. Download https://repo.maven.apache.org/maven2/commons-io/commons-io/1.4/commons-io-1.4.pomr:generateRestli > r2-1.15.7.pom
```

```
Download https://repo.maven.apache.org/maven2/com/fasterxml/jackson/core/jackson-databind/2.7.4/jackson-databind-2.7.4.pom
Download https://repo.maven.apache.org/maven2/log4j/log4j/1.2.16/log4j-1.2.16.jar
Download https://repo.maven.apache.org/maven2/com/fasterxml/jackson/module/jackson-module-scala_2.10/2.4.4/jackson-module-scala_2.10-2.4.4.jar
...(中间过程省略)
> Task:az-hadoop-jobtype-plugin:compileJava
注:某些输入文件使用或覆盖了已过时的 API。
注:有关的详细信息,请使用 -Xlint:deprecation 重新编译。
注:/usr/local/azkaban/az-hadoop-jobtype-plugin/src/main/java/azkaban/jobtype/HadoopSecureSparkWrapper.java 使用了未经检查或不安全的操作。
注:有关的详细信息,请使用 -Xlint:unchecked 重新编译。


BUILD SUCCESSFUL in 31m 14s
54 actionable tasks:54 executed
[root@YYR azkaban]#
```

注意：编译过程中可能会由于网络延时而造成编译失败，可以多试几次解决此问题。

部署 AZKABAN

④部署 Azkaban。

```
# 修改目录名
[root@YYR local]# mv azkaban/azkaban-src
# 创建 Azkaban 部署目录
[root@YYR local]# mkdir azkaban
# 将编译后的 Azkaban 部署包拷贝到部署目录
[root@YYR local]# cp azkaban-src/azkaban-db/build/distributions/azkaban-db-0.1.0-SNAPSHOT.tar.gz azkaban
[root@YYR local]# cp azkaban-src/azkaban-web-server/build/distributions/azkaban-web-server-0.1.0-SNAPSHOT.tar.gz azkaban
[root@YYR local]# cp azkaban-src/azkaban-exec-server/build/distributions/azkaban-exec-server-0.1.0-SNAPSHOT.tar.gz azkaban
```

⑤在 Azkaban 目录下解压各个编译好的压缩包，重新命名。

```
# 解压缩 azkaban-db-0.1.0-SNAPSHOT.tar.gz
[root@YYR azkaban]# tar -zxvf azkaban-db-0.1.0-SNAPSHOT.tar.gz
azkaban-db-0.1.0-SNAPSHOT/
azkaban-db-0.1.0-SNAPSHOT/create.active_executing_flows.sql
azkaban-db-0.1.0-SNAPSHOT/create.active_sla.sql
azkaban-db-0.1.0-SNAPSHOT/create.execution_dependencies.sql
azkaban-db-0.1.0-SNAPSHOT/create.execution_flows.sql
azkaban-db-0.1.0-SNAPSHOT/create.execution_jobs.sql
azkaban-db-0.1.0-SNAPSHOT/create.execution_logs.sql
azkaban-db-0.1.0-SNAPSHOT/create.executor_events.sql
azkaban-db-0.1.0-SNAPSHOT/create.executors.sql
azkaban-db-0.1.0-SNAPSHOT/create.project_events.sql
azkaban-db-0.1.0-SNAPSHOT/create.project_files.sql
azkaban-db-0.1.0-SNAPSHOT/create.project_flow_files.sql
azkaban-db-0.1.0-SNAPSHOT/create.project_flows.sql
azkaban-db-0.1.0-SNAPSHOT/create.project_permissions.sql
azkaban-db-0.1.0-SNAPSHOT/create.project_properties.sql
azkaban-db-0.1.0-SNAPSHOT/create.project_versions.sql
azkaban-db-0.1.0-SNAPSHOT/create.projects.sql
azkaban-db-0.1.0-SNAPSHOT/create.properties.sql
azkaban-db-0.1.0-SNAPSHOT/create.quartz-tables-all.sql
azkaban-db-0.1.0-SNAPSHOT/create.triggers.sql
azkaban-db-0.1.0-SNAPSHOT/database.properties
azkaban-db-0.1.0-SNAPSHOT/upgrade.3.20.0.to.3.22.0.sql
azkaban-db-0.1.0-SNAPSHOT/upgrade.3.43.0.to.3.44.0.sql
azkaban-db-0.1.0-SNAPSHOT/upgrade.3.68.0.to.3.69.0.sql
azkaban-db-0.1.0-SNAPSHOT/upgrade.3.69.0.to.3.70.0.sql
azkaban-db-0.1.0-SNAPSHOT/create-all-sql-0.1.0-SNAPSHOT.sql
# 解压缩 azkaban-exec-server-0.1.0-SNAPSHOT.tar.gz
[root@YYR azkaban]# tar -zxvf azkaban-exec-server-0.1.0-SNAPSHOT.tar.gz
azkaban-exec-server-0.1.0-SNAPSHOT/
azkaban-exec-server-0.1.0-SNAPSHOT/bin/
azkaban-exec-server-0.1.0-SNAPSHOT/bin/internal/
```

```
34. azkaban-exec-server-0.1.0-SNAPSHOT/bin/internal/internal-
    start-executor.sh
35.azkaban-exec-server-0.1.0-SNAPSHOT/bin/shutdown-exec.sh
36.azkaban-exec-server-0.1.0-SNAPSHOT/bin/start-exec.sh
37.azkaban-exec-server-0.1.0-SNAPSHOT/conf/
38.azkaban-exec-server-0.1.0-SNAPSHOT/conf/azkaban.proper-
    ties
39.azkaban-exec-server-0.1.0-SNAPSHOT/conf/global.proper-
    ties
40.azkaban-exec-server-0.1.0-SNAPSHOT/conf/log4j.properties
41.azkaban-exec-server-0.1.0-SNAPSHOT/plugins/
42.azkaban-exec-server-0.1.0-SNAPSHOT/plugins/jobtypes/
43.azkaban-exec-server-0.1.0-SNAPSHOT/plugins/jobtypes/com-
    monprivate.properties
44.azkaban-exec-server-0.1.0-SNAPSHOT/bin/internal/util.sh
45.azkaban-exec-server-0.1.0-SNAPSHOT/lib/
46.azkaban-exec-server-0.1.0-SNAPSHOT/lib/azkaban-hadoop-
    security-plugin-0.1.0-SNAPSHOT.jar
47.azkaban-exec-server-0.1.0-SNAPSHOT/lib/azkaban-common-
    0.1.0-SNAPSHOT.jar
48. azkaban-exec-server-0.1.0-SNAPSHOT/lib/azkaban-db-
    0.1.0-SNAPSHOT.jar
49. azkaban-exec-server-0.1.0-SNAPSHOT/lib/az-core-0.1.0-
    SNAPSHOT.jar
50.azkaban-exec-server-0.1.0-SNAPSHOT/lib/azkaban-spi-
    0.1.0-SNAPSHOT.jar
51.azkaban-exec-server-0.1.0-SNAPSHOT/lib/kafka-log4j-ap-
    pender-0.10.0.0.jar
52.azkaban-exec-server-0.1.0-SNAPSHOT/lib/slf4j-log4j12-
    1.7.21.jar
53.azkaban-exec-server-0.1.0-SNAPSHOT/lib/log4j-1.2.17.jar
54.azkaban-exec-server-0.1.0-SNAPSHOT/lib/guice-4.1.0.jar
55.azkaban-exec-server-0.1.0-SNAPSHOT/lib/metrics-jvm-
    3.1.0.jar
56.azkaban-exec-server-0.1.0-SNAPSHOT/lib/metrics-core-
    3.1.0.jar
57.azkaban-exec-server-0.1.0-SNAPSHOT/lib/quartz-2.2.1.jar
```

```
58.azkaban-exec-server-0.1.0-SNAPSHOT/lib/kafka-clients-0.10.0.0.jar
59.azkaban-exec-server-0.1.0-SNAPSHOT/lib/slf4j-api-1.7.21.jar
60.azkaban-exec-server-0.1.0-SNAPSHOT/lib/guava-21.0.jar
61.azkaban-exec-server-0.1.0-SNAPSHOT/lib/jsr305-3.0.2.jar
62.azkaban-exec-server-0.1.0-SNAPSHOT/lib/javax.inject-1.jar
63.azkaban-exec-server-0.1.0-SNAPSHOT/lib/aopalliance-1.0.jar
64.azkaban-exec-server-0.1.0-SNAPSHOT/lib/commons-jexl-2.1.1.jar
65.azkaban-exec-server-0.1.0-SNAPSHOT/lib/velocity-1.7.jar
66.azkaban-exec-server-0.1.0-SNAPSHOT/lib/commons-lang-2.6.jar
67.azkaban-exec-server-0.1.0-SNAPSHOT/lib/joda-time-2.0.jar
68.azkaban-exec-server-0.1.0-SNAPSHOT/lib/commons-io-2.4.jar
69.azkaban-exec-server-0.1.0-SNAPSHOT/lib/jackson-mapper-asl-1.9.5.jar
70.azkaban-exec-server-0.1.0-SNAPSHOT/lib/jackson-core-asl-1.9.5.jar
71.azkaban-exec-server-0.1.0-SNAPSHOT/lib/commons-collections-3.2.2.jar
72.azkaban-exec-server-0.1.0-SNAPSHOT/lib/commons-dbcp2-2.1.1.jar
73.azkaban-exec-server-0.1.0-SNAPSHOT/lib/commons-dbutils-1.5.jar
74.azkaban-exec-server-0.1.0-SNAPSHOT/lib/commons-fileupload-1.2.1.jar
75.azkaban-exec-server-0.1.0-SNAPSHOT/lib/gson-2.8.1.jar
76.azkaban-exec-server-0.1.0-SNAPSHOT/lib/httpclient-4.5.3.jar
77.azkaban-exec-server-0.1.0-SNAPSHOT/lib/jetty-6.1.26.jar
78.azkaban-exec-server-0.1.0-SNAPSHOT/lib/jetty-util-6.1.26.jar
```

```
azkaban-exec-server-0.1.0-SNAPSHOT/lib/jopt-simple-4.3.jar
azkaban-exec-server-0.1.0-SNAPSHOT/lib/mail-1.4.5.jar
azkaban-exec-server-0.1.0-SNAPSHOT/lib/commons-math3-3.0.jar
azkaban-exec-server-0.1.0-SNAPSHOT/lib/mysql-connector-java-5.1.28.jar
azkaban-exec-server-0.1.0-SNAPSHOT/lib/snakeyaml-1.18.jar
azkaban-exec-server-0.1.0-SNAPSHOT/lib/commons-logging-1.2.jar
azkaban-exec-server-0.1.0-SNAPSHOT/lib/c3p0-0.9.1.1.jar
azkaban-exec-server-0.1.0-SNAPSHOT/lib/commons-pool2-2.4.2.jar
azkaban-exec-server-0.1.0-SNAPSHOT/lib/httpcore-4.4.6.jar
azkaban-exec-server-0.1.0-SNAPSHOT/lib/commons-codec-1.9.jar
azkaban-exec-server-0.1.0-SNAPSHOT/lib/servlet-api-2.5-20081211.jar
azkaban-exec-server-0.1.0-SNAPSHOT/lib/activation-1.1.jar
azkaban-exec-server-0.1.0-SNAPSHOT/lib/lz4-1.3.0.jar
azkaban-exec-server-0.1.0-SNAPSHOT/lib/snappy-java-1.1.2.4.jar
azkaban-exec-server-0.1.0-SNAPSHOT/lib/azkaban-exec-server-0.1.0-SNAPSHOT.jar
#解压缩azkaban-web-server-0.1.0-SNAPSHOT.tar.gz
[root@YYR azkaban]# tar -zxvf azkaban-web-server-0.1.0-SNAPSHOT.tar.gz
azkaban-web-server-0.1.0-SNAPSHOT/
azkaban-web-server-0.1.0-SNAPSHOT/bin/
azkaban-web-server-0.1.0-SNAPSHOT/bin/internal/
azkaban-web-server-0.1.0-SNAPSHOT/bin/internal/internal-start-web.sh
azkaban-web-server-0.1.0-SNAPSHOT/bin/shutdown-web.sh
azkaban-web-server-0.1.0-SNAPSHOT/bin/start-web.sh
```

```
azkaban-web-server-0.1.0-SNAPSHOT/bin/internal/util.sh
azkaban-web-server-0.1.0-SNAPSHOT/conf/
azkaban-web-server-0.1.0-SNAPSHOT/conf/azkaban-users.xml
azkaban-web-server-0.1.0-SNAPSHOT/conf/azkaban.properties
azkaban-web-server-0.1.0-SNAPSHOT/conf/global.properties
azkaban-web-server-0.1.0-SNAPSHOT/conf/log4j.properties
azkaban-web-server-0.1.0-SNAPSHOT/lib/
azkaban-web-server-0.1.0-SNAPSHOT/lib/azkaban-common-0.1.0-SNAPSHOT.jar
azkaban-web-server-0.1.0-SNAPSHOT/lib/azkaban-db-0.1.0-SNAPSHOT.jar
azkaban-web-server-0.1.0-SNAPSHOT/lib/az-core-0.1.0-SNAPSHOT.jar
azkaban-web-server-0.1.0-SNAPSHOT/lib/az-flow-trigger-dependency-plugin-0.1.0-SNAPSHOT.jar
azkaban-web-server-0.1.0-SNAPSHOT/lib/azkaban-spi-0.1.0-SNAPSHOT.jar
azkaban-web-server-0.1.0-SNAPSHOT/lib/slf4j-log4j12-1.7.18.jar
azkaban-web-server-0.1.0-SNAPSHOT/lib/log4j-1.2.17.jar
azkaban-web-server-0.1.0-SNAPSHOT/lib/guice-4.1.0.jar
azkaban-web-server-0.1.0-SNAPSHOT/lib/restli-server-1.15.7.jar
azkaban-web-server-0.1.0-SNAPSHOT/lib/metrics-jvm-3.1.0.jar
azkaban-web-server-0.1.0-SNAPSHOT/lib/metrics-core-3.1.0.jar
azkaban-web-server-0.1.0-SNAPSHOT/lib/quartz-2.2.1.jar
azkaban-web-server-0.1.0-SNAPSHOT/lib/restli-common-1.15.7.jar
azkaban-web-server-0.1.0-SNAPSHOT/lib/li-jersey-uri-1.15.7.jar
azkaban-web-server-0.1.0-SNAPSHOT/lib/parseq-1.3.6.jar
```

124. azkaban-web-server-0.1.0-SNAPSHOT/lib/data-transform-1.15.7.jar
125. azkaban-web-server-0.1.0-SNAPSHOT/lib/r2-1.15.7.jar
126. azkaban-web-server-0.1.0-SNAPSHOT/lib/data-1.15.7.jar
127. azkaban-web-server-0.1.0-SNAPSHOT/lib/pegasus-common-1.15.7.jar
128. azkaban-web-server-0.1.0-SNAPSHOT/lib/mina-core-1.1.7.jar
129. azkaban-web-server-0.1.0-SNAPSHOT/lib/slf4j-api-1.7.18.jar
130. azkaban-web-server-0.1.0-SNAPSHOT/lib/velocity-tools-2.0.jar
131. azkaban-web-server-0.1.0-SNAPSHOT/lib/javax.inject-1.jar
132. azkaban-web-server-0.1.0-SNAPSHOT/lib/aopalliance-1.0.jar
133. azkaban-web-server-0.1.0-SNAPSHOT/lib/guava-21.0.jar
134. azkaban-web-server-0.1.0-SNAPSHOT/lib/commons-jexl-2.1.1.jar
135. azkaban-web-server-0.1.0-SNAPSHOT/lib/velocity-1.7.jar
136. azkaban-web-server-0.1.0-SNAPSHOT/lib/commons-lang-2.6.jar
137. azkaban-web-server-0.1.0-SNAPSHOT/lib/joda-time-2.0.jar
138. azkaban-web-server-0.1.0-SNAPSHOT/lib/commons-io-2.4.jar
139. azkaban-web-server-0.1.0-SNAPSHOT/lib/jackson-mapper-asl-1.9.5.jar
140. azkaban-web-server-0.1.0-SNAPSHOT/lib/jackson-core-asl-1.9.5.jar
141. azkaban-web-server-0.1.0-SNAPSHOT/lib/commons-collections-3.2.2.jar
142. azkaban-web-server-0.1.0-SNAPSHOT/lib/commons-dbcp2-2.1.1.jar
143. azkaban-web-server-0.1.0-SNAPSHOT/lib/commons-dbutils-1.5.jar

```
144. azkaban-web-server-0.1.0-SNAPSHOT/lib/commons-fileup-
     load-1.2.1.jar
145.azkaban-web-server-0.1.0-SNAPSHOT/lib/gson-2.8.1.jar
146.azkaban-web-server-0.1.0-SNAPSHOT/lib/httpclient-
     4.5.3.jar
147.azkaban-web-server-0.1.0-SNAPSHOT/lib/jetty-6.1.26.jar
148.azkaban-web-server-0.1.0-SNAPSHOT/lib/jetty-util-
     6.1.26.jar
149.azkaban-web-server-0.1.0-SNAPSHOT/lib/jopt-simple-
     4.3.jar
150.azkaban-web-server-0.1.0-SNAPSHOT/lib/mail-1.4.5.jar
151.azkaban-web-server-0.1.0-SNAPSHOT/lib/commons-math3-
     3.0.jar
152.azkaban-web-server-0.1.0-SNAPSHOT/lib/mysql-connector-
     java-5.1.28.jar
153.azkaban-web-server-0.1.0-SNAPSHOT/lib/snakeyaml-
     1.18.jar
154.azkaban-web-server-0.1.0-SNAPSHOT/lib/javax.servlet-
     api-3.0.1.jar
155. azkaban-web-server-0.1.0-SNAPSHOT/lib/jackson-core-
     2.2.2.jar
156.azkaban-web-server-0.1.0-SNAPSHOT/lib/struts-taglib-
     1.3.8.jar
157.azkaban-web-server-0.1.0-SNAPSHOT/lib/struts-tiles-
     1.3.8.jar
158.azkaban-web-server-0.1.0-SNAPSHOT/lib/struts-core-
     1.3.8.jar
159.azkaban-web-server-0.1.0-SNAPSHOT/lib/commons-chain-
     1.1.jar
160.azkaban-web-server-0.1.0-SNAPSHOT/lib/commons-valida-
     tor-1.3.1.jar
161.azkaban-web-server-0.1.0-SNAPSHOT/lib/commons-digester-
     1.8.jar
162.azkaban-web-server-0.1.0-SNAPSHOT/lib/commons-beanu-
     tils-1.7.0.jar
163.azkaban-web-server-0.1.0-SNAPSHOT/lib/commons-logging-
     1.2.jar
```

```
164. azkaban-web-server-0.1.0-SNAPSHOT/lib/dom4j-1.1.jar
165. azkaban-web-server-0.1.0-SNAPSHOT/lib/oro-2.0.8.jar
166. azkaban-web-server-0.1.0-SNAPSHOT/lib/sslext-1.2-0.jar
167. azkaban-web-server-0.1.0-SNAPSHOT/lib/c3p0-0.9.1.1.jar
168. azkaban-web-server-0.1.0-SNAPSHOT/lib/commons-pool2-2.4.2.jar
169. azkaban-web-server-0.1.0-SNAPSHOT/lib/httpcore-4.4.6.jar
170. azkaban-web-server-0.1.0-SNAPSHOT/lib/commons-codec-1.9.jar
171. azkaban-web-server-0.1.0-SNAPSHOT/lib/servlet-api-2.5-20081211.jar
172. azkaban-web-server-0.1.0-SNAPSHOT/lib/json-20070829.jar
173. azkaban-web-server-0.1.0-SNAPSHOT/lib/cglib-nodep-2.2.jar
174. azkaban-web-server-0.1.0-SNAPSHOT/lib/netty-3.2.3.Final.jar
175. azkaban-web-server-0.1.0-SNAPSHOT/lib/commons-compress-1.2.jar
176. azkaban-web-server-0.1.0-SNAPSHOT/lib/snappy-0.3.jar
177. azkaban-web-server-0.1.0-SNAPSHOT/lib/antlr-2.7.2.jar
178. azkaban-web-server-0.1.0-SNAPSHOT/lib/activation-1.1.jar
179. azkaban-web-server-0.1.0-SNAPSHOT/lib/azkaban-web-server-0.1.0-SNAPSHOT.jar
180. azkaban-web-server-0.1.0-SNAPSHOT/web/
181. azkaban-web-server-0.1.0-SNAPSHOT/web/css/
182. azkaban-web-server-0.1.0-SNAPSHOT/web/css/bootstrap-datetimepicker.css
183. azkaban-web-server-0.1.0-SNAPSHOT/web/css/bootstrap.css
184. azkaban-web-server-0.1.0-SNAPSHOT/web/css/images/
185. azkaban-web-server-0.1.0-SNAPSHOT/web/css/images/addIcon.png
```

```
186. azkaban-web-server-0.1.0-SNAPSHOT/web/css/images/anima-
    ted-overlay.gif
187. azkaban-web-server-0.1.0-SNAPSHOT/web/css/images/dot-
    icon.png
188. azkaban-web-server-0.1.0-SNAPSHOT/web/css/images/red
    warning.png
189. azkaban-web-server-0.1.0-SNAPSHOT/web/css/images/re-
    moveIcon.png
190. azkaban-web-server-0.1.0-SNAPSHOT/web/css/images/ui-bg_
    flat_0_aaaaaa_40x100.png
191. azkaban-web-server-0.1.0-SNAPSHOT/web/css/images/ui-bg_
    flat_0_eeeeee_40x100.png
192. azkaban-web-server-0.1.0-SNAPSHOT/web/css/images/ui-bg_
    flat_35_dddddd_40x100.png
193. azkaban-web-server-0.1.0-SNAPSHOT/web/css/images/ui-bg_
    flat_55_c0402a_40x100.png
194. azkaban-web-server-0.1.0-SNAPSHOT/web/css/images/ui-bg_
    flat_55_eeeeee_40x100.png
195. azkaban-web-server-0.1.0-SNAPSHOT/web/css/images/ui-bg_
    flat_75_ffffff_40x100.png
196. azkaban-web-server-0.1.0-SNAPSHOT/web/css/images/ui-bg_
    glass_100_f8f8f8_1x400.png
197. azkaban-web-server-0.1.0-SNAPSHOT/web/css/images/ui-bg_
    glass_55_fbf9ee_1x400.png
198. azkaban-web-server-0.1.0-SNAPSHOT/web/css/images/ui-bg_
    glass_60_eeeeee_1x400.png
199. azkaban-web-server-0.1.0-SNAPSHOT/web/css/images/ui-bg_
    glass_65_ffffff_1x400.png
200. azkaban-web-server-0.1.0-SNAPSHOT/web/css/images/ui-bg_
    glass_75_dadada_1x400.png
201. azkaban-web-server-0.1.0-SNAPSHOT/web/css/images/ui-bg_
    glass_75_e6e6e6_1x400.png
202. azkaban-web-server-0.1.0-SNAPSHOT/web/css/images/ui-bg_
    glass_95_fef1ec_1x400.png
203. azkaban-web-server-0.1.0-SNAPSHOT/web/css/images/ui-bg_
    highlight-soft_75_cccccc_1x100.png
```

204. azkaban-web-server-0.1.0-SNAPSHOT/web/css/images/ui-bg_inset-hard_75_999999_1x100.png
205. azkaban-web-server-0.1.0-SNAPSHOT/web/css/images/ui-bg_inset-soft_50_c9c9c9_1x100.png
206. azkaban-web-server-0.1.0-SNAPSHOT/web/css/images/ui-icons_000000_256x240.png
207. azkaban-web-server-0.1.0-SNAPSHOT/web/css/images/ui-icons_222222_256x240.png
208. azkaban-web-server-0.1.0-SNAPSHOT/web/css/images/ui-icons_2e83ff_256x240.png
209. azkaban-web-server-0.1.0-SNAPSHOT/web/css/images/ui-icons_3383bb_256x240.png
210. azkaban-web-server-0.1.0-SNAPSHOT/web/css/images/ui-icons_454545_256x240.png
211. azkaban-web-server-0.1.0-SNAPSHOT/web/css/images/ui-icons_70b2e1_256x240.png
212. azkaban-web-server-0.1.0-SNAPSHOT/web/css/images/ui-icons_888888_256x240.png
213. azkaban-web-server-0.1.0-SNAPSHOT/web/css/images/ui-icons_999999_256x240.png
214. azkaban-web-server-0.1.0-SNAPSHOT/web/css/images/ui-icons_cccccc_256x240.png
215. azkaban-web-server-0.1.0-SNAPSHOT/web/css/images/ui-icons_cd0a0a_256x240.png
216. azkaban-web-server-0.1.0-SNAPSHOT/web/css/images/ui-icons_fbc856_256x240.png
217. azkaban-web-server-0.1.0-SNAPSHOT/web/css/images/ui-icons_ffffff_256x240.png
218. azkaban-web-server-0.1.0-SNAPSHOT/web/css/jquery-ui-1.10.1.custom.css
219. azkaban-web-server-0.1.0-SNAPSHOT/web/css/jquery-ui-timepicker-addon.css
220. azkaban-web-server-0.1.0-SNAPSHOT/web/css/jquery-ui.css
221. azkaban-web-server-0.1.0-SNAPSHOT/web/css/jquery.svg.css
222. azkaban-web-server-0.1.0-SNAPSHOT/web/css/morris.css

```
223. azkaban-web-server-0.1.0-SNAPSHOT/web/favicon.ico
224. azkaban-web-server-0.1.0-SNAPSHOT/web/fonts/
225. azkaban-web-server-0.1.0-SNAPSHOT/web/fonts/glyphicons-
     halflings-regular.eot
226. azkaban-web-server-0.1.0-SNAPSHOT/web/fonts/glyphicons-
     halflings-regular.svg
227. azkaban-web-server-0.1.0-SNAPSHOT/web/fonts/glyphicons-
     halflings-regular.ttf
228. azkaban-web-server-0.1.0-SNAPSHOT/web/fonts/glyphicons-
     halflings-regular.woff
229. azkaban-web-server-0.1.0-SNAPSHOT/web/images/
230. azkaban-web-server-0.1.0-SNAPSHOT/web/images/graph-
     icon.png
231. azkaban-web-server-0.1.0-SNAPSHOT/web/images/logo.png
232. azkaban-web-server-0.1.0-SNAPSHOT/web/images/warn-
     ing.png
233. azkaban-web-server-0.1.0-SNAPSHOT/web/js/
234. azkaban-web-server-0.1.0-SNAPSHOT/web/js/azkaban/
235. azkaban-web-server-0.1.0-SNAPSHOT/web/js/azkaban/model/
236. azkaban-web-server-0.1.0-SNAPSHOT/web/js/azkaban/mod-
     el/flow-trigger.js
237. azkaban-web-server-0.1.0-SNAPSHOT/web/js/azkaban/model/
     job-log.js
238. azkaban-web-server-0.1.0-SNAPSHOT/web/js/azkaban/model/
     svg-graph.js
239. azkaban-web-server-0.1.0-SNAPSHOT/web/js/azkaban/
     namespace.js
240. azkaban-web-server-0.1.0-SNAPSHOT/web/js/azkaban/test/
241. azkaban-web-server-0.1.0-SNAPSHOT/web/js/azkaban/test/
     test.js
242. azkaban-web-server-0.1.0-SNAPSHOT/web/js/azkaban/util/
243. azkaban-web-server-0.1.0-SNAPSHOT/web/js/azkaban/util/
     ajax.js
244. azkaban-web-server-0.1.0-SNAPSHOT/web/js/azkaban/util/
     common.js
245. azkaban-web-server-0.1.0-SNAPSHOT/web/js/azkaban/util/
     date.js
```

246. azkaban-web-server-0.1.0-SNAPSHOT/web/js/azkaban/util/flow-loader.js
247. azkaban-web-server-0.1.0-SNAPSHOT/web/js/azkaban/util/job-status.js
248. azkaban-web-server-0.1.0-SNAPSHOT/web/js/azkaban/util/layout.js
249. azkaban-web-server-0.1.0-SNAPSHOT/web/js/azkaban/util/schedule.js
250. azkaban-web-server-0.1.0-SNAPSHOT/web/js/azkaban/util/svg-navigate.js
251. azkaban-web-server-0.1.0-SNAPSHOT/web/js/azkaban/util/svgutils.js
252. azkaban-web-server-0.1.0-SNAPSHOT/web/js/azkaban/view/
253. azkaban-web-server-0.1.0-SNAPSHOT/web/js/azkaban/view/admin-setup.js
254. azkaban-web-server-0.1.0-SNAPSHOT/web/js/azkaban/view/context-menu.js
255. azkaban-web-server-0.1.0-SNAPSHOT/web/js/azkaban/view/executions.js
256. azkaban-web-server-0.1.0-SNAPSHOT/web/js/azkaban/view/exflow.js
257. azkaban-web-server-0.1.0-SNAPSHOT/web/js/azkaban/view/flow-execute-dialog.js
258. azkaban-web-server-0.1.0-SNAPSHOT/web/js/azkaban/view/flow-execution-list.js
259. azkaban-web-server-0.1.0-SNAPSHOT/web/js/azkaban/view/flow-extended.js
260. azkaban-web-server-0.1.0-SNAPSHOT/web/js/azkaban/view/flow-stats.js
261. azkaban-web-server-0.1.0-SNAPSHOT/web/js/azkaban/view/flow-trigger-list.js
262. azkaban-web-server-0.1.0-SNAPSHOT/web/js/azkaban/view/flow.js
263. azkaban-web-server-0.1.0-SNAPSHOT/web/js/azkaban/view/history-day.js
264. azkaban-web-server-0.1.0-SNAPSHOT/web/js/azkaban/view/history.js

```
265. azkaban-web-server-0.1.0-SNAPSHOT/web/js/azkaban/view/
     jmx.js
266. azkaban-web-server-0.1.0-SNAPSHOT/web/js/azkaban/view/
     job-details.js
267. azkaban-web-server-0.1.0-SNAPSHOT/web/js/azkaban/view/
     job-edit.js
268. azkaban-web-server-0.1.0-SNAPSHOT/web/js/azkaban/view/
     job-history.js
269. azkaban-web-server-0.1.0-SNAPSHOT/web/js/azkaban/view/
     job-list.js
270. azkaban-web-server-0.1.0-SNAPSHOT/web/js/azkaban/view/
     login.js
271. azkaban-web-server-0.1.0-SNAPSHOT/web/js/azkaban/view/
     main.js
272. azkaban-web-server-0.1.0-SNAPSHOT/web/js/azkaban/view/
     message-dialog.js
273. azkaban-web-server-0.1.0-SNAPSHOT/web/js/azkaban/view/
     note.js
274. azkaban-web-server-0.1.0-SNAPSHOT/web/js/azkaban/view/
     project-logs.js
275. azkaban-web-server-0.1.0-SNAPSHOT/web/js/azkaban/view/
     project-modals.js
276. azkaban-web-server-0.1.0-SNAPSHOT/web/js/azkaban/view/
     project-permissions.js
277. azkaban-web-server-0.1.0-SNAPSHOT/web/js/azkaban/view/
     project.js
278. azkaban-web-server-0.1.0-SNAPSHOT/web/js/azkaban/view/
     schedule-options.js
279. azkaban-web-server-0.1.0-SNAPSHOT/web/js/azkaban/view/
     schedule-panel.js
280. azkaban-web-server-0.1.0-SNAPSHOT/web/js/azkaban/view/
     schedule-sla.js
281. azkaban-web-server-0.1.0-SNAPSHOT/web/js/azkaban/view/
     scheduled.js
282. azkaban-web-server-0.1.0-SNAPSHOT/web/js/azkaban/view/
     svg-graph.js
```

283. azkaban-web-server-0.1.0-SNAPSHOT/web/js/azkaban/view/table-sort.js
284. azkaban-web-server-0.1.0-SNAPSHOT/web/js/azkaban/view/time-graph.js
285. azkaban-web-server-0.1.0-SNAPSHOT/web/js/azkaban/view/triggers.js
286. azkaban-web-server-0.1.0-SNAPSHOT/web/js/backbone-0.9.10-min.js
287. azkaban-web-server-0.1.0-SNAPSHOT/web/js/bootstrap.min.js
288. azkaban-web-server-0.1.0-SNAPSHOT/web/js/d3.v3.min.js
289. azkaban-web-server-0.1.0-SNAPSHOT/web/js/dust-full-2.2.3.min.js
290. azkaban-web-server-0.1.0-SNAPSHOT/web/js/jquery.svg.js
291. azkaban-web-server-0.1.0-SNAPSHOT/web/js/jquery.svg.min.js
292. azkaban-web-server-0.1.0-SNAPSHOT/web/js/jquery.svganim.min.js
293. azkaban-web-server-0.1.0-SNAPSHOT/web/js/jquery.svgdom.js
294. azkaban-web-server-0.1.0-SNAPSHOT/web/js/jquery.svgdom.min.js
295. azkaban-web-server-0.1.0-SNAPSHOT/web/js/jquery.svgfilter.min.js
296. azkaban-web-server-0.1.0-SNAPSHOT/web/js/jquery.twbsPagination.min.js
297. azkaban-web-server-0.1.0-SNAPSHOT/web/js/jquery/
298. azkaban-web-server-0.1.0-SNAPSHOT/web/js/jquery/icons/
299. azkaban-web-server-0.1.0-SNAPSHOT/web/js/jquery/icons/file.png
300. azkaban-web-server-0.1.0-SNAPSHOT/web/js/jquery/icons/flow.png
301. azkaban-web-server-0.1.0-SNAPSHOT/web/js/jquery/icons/folder.png
302. azkaban-web-server-0.1.0-SNAPSHOT/web/js/jquery/icons/folderopen.png

```
303. azkaban-web-server-0.1.0-SNAPSHOT/web/js/jquery/icons/job.png
304. azkaban-web-server-0.1.0-SNAPSHOT/web/js/jquery/jquery-1.9.1.js
305. azkaban-web-server-0.1.0-SNAPSHOT/web/js/jquery/jquery.autocomplete.css
306. azkaban-web-server-0.1.0-SNAPSHOT/web/js/jquery/jquery.autocomplete.pack.js
307. azkaban-web-server-0.1.0-SNAPSHOT/web/js/jquery/jquery.contextMenu.css
308. azkaban-web-server-0.1.0-SNAPSHOT/web/js/jquery/jquery.contextMenu.js
309. azkaban-web-server-0.1.0-SNAPSHOT/web/js/jquery/jquery.cookie.js
310. azkaban-web-server-0.1.0-SNAPSHOT/web/js/jquery/jquery.hotkeys.js
311. azkaban-web-server-0.1.0-SNAPSHOT/web/js/jquery/jquery.js
312. azkaban-web-server-0.1.0-SNAPSHOT/web/js/jquery/jquery.jstree.js
313. azkaban-web-server-0.1.0-SNAPSHOT/web/js/jquery/jquery.svg.min.js
314. azkaban-web-server-0.1.0-SNAPSHOT/web/js/jquery/jquery.tablesorter.js
315. azkaban-web-server-0.1.0-SNAPSHOT/web/js/jquery/jquery.tablesorter.min.js
316. azkaban-web-server-0.1.0-SNAPSHOT/web/js/jquery/jquery.tools.min.js
317. azkaban-web-server-0.1.0-SNAPSHOT/web/js/jquery/jquery.treeTable.min.js
318. azkaban-web-server-0.1.0-SNAPSHOT/web/js/jquery/themes/
319. azkaban-web-server-0.1.0-SNAPSHOT/web/js/jquery/themes/apple/
320. azkaban-web-server-0.1.0-SNAPSHOT/web/js/jquery/themes/apple/bg.jpg
321. azkaban-web-server-0.1.0-SNAPSHOT/web/js/jquery/themes/apple/d.png
```

322. azkaban - web - server - 0.1.0 - SNAPSHOT/ web/ js/ jquery/ themes/apple/dot_for_ie.gif
323. azkaban - web - server - 0.1.0 - SNAPSHOT/web/js/jquery/themes/apple/style.css
324. azkaban - web - server - 0.1.0 - SNAPSHOT/web/js/jquery/themes/apple/throbber.gif
325. azkaban - web - server - 0.1.0 - SNAPSHOT/web/js/jquery/themes/classic/
326. azkaban - web - server - 0.1.0 - SNAPSHOT/web/js/jquery/themes/classic/d.png
327. azkaban - web - server - 0.1.0 - SNAPSHOT/web/js/jquery/themes/classic/dot_for_ie.gif
328. azkaban - web - server - 0.1.0 - SNAPSHOT/web/js/jquery/themes/classic/style.css
329. azkaban - web - server - 0.1.0 - SNAPSHOT/web/js/jquery/themes/classic/throbber.gif
330. azkaban - web - server - 0.1.0 - SNAPSHOT/web/js/jquery/themes/default - rtl/
331. azkaban - web - server - 0.1.0 - SNAPSHOT/ web/ js/ jquery/ themes/default - rtl/d.gif
332. azkaban - web - server - 0.1.0 - SNAPSHOT/web/js/jquery/themes/default - rtl/d.png
333. azkaban - web - server - 0.1.0 - SNAPSHOT/web/js/jquery/themes/default - rtl/dots.gif
334. azkaban - web - server - 0.1.0 - SNAPSHOT/web/js/jquery/themes/default - rtl/style.css
335. azkaban - web - server - 0.1.0 - SNAPSHOT/web/js/jquery/themes/default - rtl/throbber.gif
336. azkaban - web - server - 0.1.0 - SNAPSHOT/web/js/jquery/themes/default/
337. azkaban - web - server - 0.1.0 - SNAPSHOT/web/js/jquery/themes/default/d.gif
338. azkaban - web - server - 0.1.0 - SNAPSHOT/web/js/jquery/themes/default/d.png
339. azkaban - web - server - 0.1.0 - SNAPSHOT/web/js/jquery/themes/default/style.css

```
340. azkaban-web-server-0.1.0-SNAPSHOT/web/js/jquery/themes/default/throbber.gif
341. azkaban-web-server-0.1.0-SNAPSHOT/web/js/jqueryui/
342. azkaban-web-server-0.1.0-SNAPSHOT/web/js/jqueryui/jquery-ui-1.10.1.custom.js
343. azkaban-web-server-0.1.0-SNAPSHOT/web/js/jqueryui/jquery-ui-sliderAccess.js
344. azkaban-web-server-0.1.0-SNAPSHOT/web/js/jqueryui/jquery-ui-timepicker-addon.js
345. azkaban-web-server-0.1.0-SNAPSHOT/web/js/jqueryui/jquery.effects.blind.min.js
346. azkaban-web-server-0.1.0-SNAPSHOT/web/js/jqueryui/jquery.effects.bounce.min.js
347. azkaban-web-server-0.1.0-SNAPSHOT/web/js/jqueryui/jquery.effects.clip.min.js
348. azkaban-web-server-0.1.0-SNAPSHOT/web/js/jqueryui/jquery.effects.core.min.js
349. azkaban-web-server-0.1.0-SNAPSHOT/web/js/jqueryui/jquery.effects.drop.min.js
350. azkaban-web-server-0.1.0-SNAPSHOT/web/js/jqueryui/jquery.effects.explode.min.js
351. azkaban-web-server-0.1.0-SNAPSHOT/web/js/jqueryui/jquery.effects.fade.min.js
352. azkaban-web-server-0.1.0-SNAPSHOT/web/js/jqueryui/jquery.effects.fold.min.js
353. azkaban-web-server-0.1.0-SNAPSHOT/web/js/jqueryui/jquery.effects.highlight.min.js
354. azkaban-web-server-0.1.0-SNAPSHOT/web/js/jqueryui/jquery.effects.pulsate.min.js
355. azkaban-web-server-0.1.0-SNAPSHOT/web/js/jqueryui/jquery.effects.scale.min.js
356. azkaban-web-server-0.1.0-SNAPSHOT/web/js/jqueryui/jquery.effects.shake.min.js
357. azkaban-web-server-0.1.0-SNAPSHOT/web/js/jqueryui/jquery.effects.slide.min.js
358. azkaban-web-server-0.1.0-SNAPSHOT/web/js/jqueryui/jquery.effects.transfer.min.js
```

359. azkaban - web - server - 0.1.0 - SNAPSHOT/web/js/jqueryui/jquery.ui.accordion.min.js
360. azkaban - web - server - 0.1.0 - SNAPSHOT/web/js/jqueryui/jquery.ui.autocomplete.min.js
361. azkaban - web - server - 0.1.0 - SNAPSHOT/web/js/jqueryui/jquery.ui.button.min.js
362. azkaban - web - server - 0.1.0 - SNAPSHOT/web/js/jqueryui/jquery.ui.core.min.js
363. azkaban - web - server - 0.1.0 - SNAPSHOT/web/js/jqueryui/jquery.ui.datepicker.min.js
364. azkaban - web - server - 0.1.0 - SNAPSHOT/web/js/jqueryui/jquery.ui.dialog.min.js
365. azkaban - web - server - 0.1.0 - SNAPSHOT/web/js/jqueryui/jquery.ui.draggable.min.js
366. azkaban - web - server - 0.1.0 - SNAPSHOT/web/js/jqueryui/jquery.ui.droppable.min.js
367. azkaban - web - server - 0.1.0 - SNAPSHOT/web/js/jqueryui/jquery.ui.mouse.min.js
368. azkaban - web - server - 0.1.0 - SNAPSHOT/web/js/jqueryui/jquery.ui.position.min.js
369. azkaban - web - server - 0.1.0 - SNAPSHOT/web/js/jqueryui/jquery.ui.progressbar.min.js
370. azkaban - web - server - 0.1.0 - SNAPSHOT/web/js/jqueryui/jquery.ui.resizable.min.js
371. azkaban - web - server - 0.1.0 - SNAPSHOT/web/js/jqueryui/jquery.ui.selectable.min.js
372. azkaban - web - server - 0.1.0 - SNAPSHOT/web/js/jqueryui/jquery.ui.slider.min.js
373. azkaban - web - server - 0.1.0 - SNAPSHOT/web/js/jqueryui/jquery.ui.sortable.min.js
374. azkaban - web - server - 0.1.0 - SNAPSHOT/web/js/jqueryui/jquery.ui.tabs.min.js
375. azkaban - web - server - 0.1.0 - SNAPSHOT/web/js/jqueryui/jquery.ui.widget.min.js
376. azkaban - web - server - 0.1.0 - SNAPSHOT/web/js/morris.min.js
377. azkaban - web - server - 0.1.0 - SNAPSHOT/web/js/raphael.min.js

```
azkaban-web-server-0.1.0-SNAPSHOT/web/js/underscore-1.4.4-min.js
azkaban-web-server-0.1.0-SNAPSHOT/web/css/azkaban-graph.css
azkaban-web-server-0.1.0-SNAPSHOT/web/css/azkaban.css
azkaban-web-server-0.1.0-SNAPSHOT/web/js/flowstats-no-data.js
azkaban-web-server-0.1.0-SNAPSHOT/web/js/flowstats.js
azkaban-web-server-0.1.0-SNAPSHOT/web/js/flowsummary.js
azkaban-web-server-0.1.0-SNAPSHOT/web/js/later.min.js
azkaban-web-server-0.1.0-SNAPSHOT/web/js/moment.min.js
azkaban-web-server-0.1.0-SNAPSHOT/web/js/bootstrap-datetimepicker.min.js
azkaban-web-server-0.1.0-SNAPSHOT/web/js/moment-timezone-with-data.min.js
#重命名
[root@YYR azkaban]# mv azkaban-db-0.1.0-SNAPSHOT azkaban-db
#重命名
[root@YYR azkaban]# mv azkaban-exec-server-0.1.0-SNAPSHOT azkaban-exec
#重命名
[root@YYR azkaban]# mv azkaban-web-server-0.1.0-SNAPSHOT azkaban-web
[root@YYR azkaban]#
```

通过一系列的操作，Azkaban 安装完成，接下来配置 Azkaban，使其能够运行。

导入数据库

任务2　导入数据库 

在 azkaban-db 中有基本的数据库，需要将数据导入 MySQL 数据库中。MySQL 数据库可以和当前 Azkaban 安装在同一节点上，也可以安装在不同的节点上，本任务中 Azkaban 安装在 mynode5 节点上，MySQL 数据库安装在 mynode2 节点上。

1. 登录 MySQL 数据库，创建 Azkaban 数据库

```
#登录 MySQL
[root@YYR azkaban]# mysql -u root -p
```

```
Enter password:
Welcome to the MySQL monitor. Commands end with;or \g.
Your MySQL connection id is 3443
Server version:5.7.26 MySQL Community Server (GPL)

Copyright (c) 2000,2019,Oracle and/or its affiliates. All rights reserved.

Oracle is a registered trademark of Oracle Corporation and/or its
affiliates. Other names may be trademarks of their respective
owners.

Type 'help;' or '\h' for help. Type '\c' to clear the current input statement.

# 创建 Azkaban 数据库
mysql >create database azkaban;
Query OK,1 row affected (0.00 sec)

# 查看数据创建是否成功
mysql >show databases;
+--------------------+
| Database           |
+--------------------+
| information_schema |
| azkaban            |
| hive               |
| my_keyword         |
| mysql              |
| performance_schema |
| sys                |
+--------------------+
7 rows in set (0.01 sec)

# 退出 MySQL
mysql >exit;
Bye
```

2. 向 MySQL 中导入数据库

```
# 登录 MySQL
[root@YYR azkaban]# mysql -u root -p
Enter password:
Welcome to the MySQL monitor.  Commands end with;or \g.
Your MySQL connection id is 3445
Server version:5.7.26 MySQL Community Server (GPL)
Copyright (c) 2000,2019,Oracle and/or its affiliates. All rights reserved.
Oracle is a registered trademark of Oracle Corporation and/or its
affiliates. Other names may be trademarks of their respective
owners.
Type 'help;' or '\h' for help. Type '\c' to clear the current input statement.
# 切换当前数据库
mysql >use azkaban;
Database changed
# 导入 Azkaban 数据表
mysql > source /usr/local/azkaban/azkaban-db/create-all-sql-0.1.0-SNAPSHOT.sql
Query OK,0 rows affected (0.65 sec)
Query OK,0 rows affected (0.10 sec)
Query OK,0 rows affected (0.07 sec)
Query OK,0 rows affected (0.02 sec)
Records:0  Duplicates:0  Warnings:0
Query OK,0 rows affected (0.01 sec)
```

```
Query OK,0 rows affected (0.05 sec)
Records:0  Duplicates:0  Warnings:0
Query OK,0 rows affected (0.04 sec)
Records:0  Duplicates:0  Warnings:0
Query OK,0 rows affected (0.06 sec)
Records:0  Duplicates:0  Warnings:0
Query OK,0 rows affected (0.05 sec)
Records:0  Duplicates:0  Warnings:0
Query OK,0 rows affected (0.06 sec)
Records:0  Duplicates:0  Warnings:0
Query OK,0 rows affected (0.01 sec)
Records:0  Duplicates:0  Warnings:0
Query OK,0 rows affected (0.04 sec)
Query OK,0 rows affected (0.01 sec)
Records:0  Duplicates:0  Warnings:0
Query OK,0 rows affected (0.13 sec)
Query OK,0 rows affected (0.01 sec)
Records:0  Duplicates:0  Warnings:0
Query OK,0 rows affected (0.04 sec)
Records:0  Duplicates:0  Warnings:0
Query OK,0 rows affected (0.04 sec)
Records:0  Duplicates:0  Warnings:0
Query OK,0 rows affected (0.06 sec)
```

```
Query OK,0 rows affected (0.05 sec)
Records:0  Duplicates:0  Warnings:0
Query OK,0 rows affected (0.02 sec)
Query OK,0 rows affected (0.03 sec)
Records:0  Duplicates:0  Warnings:0
Query OK,0 rows affected (0.08 sec)
Query OK,0 rows affected (0.01 sec)
Records:0  Duplicates:0  Warnings:0
Query OK,0 rows affected (0.08 sec)
Query OK,0 rows affected (0.06 sec)
Records:0  Duplicates:0  Warnings:0
Query OK,0 rows affected (0.04 sec)
Query OK,0 rows affected (0.11 sec)
Query OK,0 rows affected (0.01 sec)
Records:0  Duplicates:0  Warnings:0
Query OK,0 rows affected (0.01 sec)
Query OK,0 rows affected (0.02 sec)
Records:0  Duplicates:0  Warnings:0
Query OK,0 rows affected (0.10 sec)
Query OK,0 rows affected (0.04 sec)
Records:0  Duplicates:0  Warnings:0
Query OK,0 rows affected (0.02 sec)
```

```
Query OK,0 rows affected (0.08 sec)
Records:0  Duplicates:0  Warnings:0
Query OK,0 rows affected (0.04 sec)
Query OK,0 rows affected (0.19 sec)
Records:0  Duplicates:0  Warnings:0
Query OK,0 rows affected (0.03 sec)
Query OK,0 rows affected,1 warning (0.00 sec)
Query OK,0 rows affected,1 warning (0.00 sec)
Query OK,0 rows affected,1 warning (0.00 sec)
Query OK,0 rows affected,1 warning (0.00 sec)
Query OK,0 rows affected,1 warning (0.00 sec)
Query OK,0 rows affected,1 warning (0.00 sec)
Query OK,0 rows affected,1 warning (0.00 sec)
Query OK,0 rows affected,1 warning (0.00 sec)
Query OK,0 rows affected,1 warning (0.00 sec)
Query OK,0 rows affected,1 warning (0.00 sec)
Query OK,0 rows affected,1 warning (0.00 sec)
Query OK,0 rows affected (0.04 sec)
Query OK,0 rows affected (0.06 sec)
Query OK,0 rows affected (0.05 sec)
```

```
Query OK,0 rows affected (0.04 sec)
Query OK,0 rows affected (0.05 sec)
Query OK,0 rows affected (0.11 sec)
Query OK,0 rows affected (0.02 sec)
Query OK,0 rows affected (0.15 sec)
Query OK,0 rows affected (0.01 sec)
Query OK,0 rows affected (0.25 sec)
Query OK,0 rows affected (0.04 sec)
Query OK,0 rows affected (0.00 sec)
Query OK,0 rows affected (0.01 sec)
```

3. 检查导入的数据库表

```
# 查看数据表导入是否正确
mysql > show tables;
+---------------------------+
| Tables_in_azkaban         |
+---------------------------+
| QRTZ_BLOB_TRIGGERS        |
| QRTZ_CALENDARS            |
| QRTZ_CRON_TRIGGERS        |
| QRTZ_FIRED_TRIGGERS       |
| QRTZ_JOB_DETAILS          |
| QRTZ_LOCKS                |
| QRTZ_PAUSED_TRIGGER_GRPS  |
| QRTZ_SCHEDULER_STATE      |
| QRTZ_SIMPLE_TRIGGERS      |
```

```
|QRTZ_SIMPROP_TRIGGERS          |
|QRTZ_TRIGGERS                  |
|active_executing_flows         |
|active_sla                     |
|execution_dependencies         |
|execution_flows                |
|execution_jobs                 |
|execution_logs                 |
|executor_events                |
|executors                      |
|project_events                 |
|project_files                  |
|project_flow_files             |
|project_flows                  |
|project_permissions            |
|project_properties             |
|project_versions               |
|projects                       |
|properties                     |
|triggers                       |
+-------------------------------+
29 rows in set (0.00 sec)
```

(1) 创建SSL配置

```
# 在 Azkaban 目录下生成 SSL 密钥
[root@YYR azkaban]# keytool -keystore keystore -alias jetty -genkey -keyalg RSA
输入密钥库口令:(此处输入000000)
再次输入新口令:(此处输入000000)
您的名字与姓氏是什么?
[Unknown]:azkaban
您的组织的单位名称是什么?
[Unknown]:azkaban
您的组织名称是什么?
[Unknown]:azkaban
您所在的城市或区域名称是什么?
[Unknown]:beijing
```

创建SSL配置

```
您所在的省/市/自治区名称是什么?
[Unknown]:beijing
该单位的双字母国家/地区代码是什么?
[Unknown]:CN
CN = azkaban,OU = azkaban,O = azkaban,L = beijing,ST = beijing,C = CN 是否正确?
[否]:y
输入 <jetty> 的密钥口令
        (如果和密钥库口令相同,按 Enter 键):
再次输入新口令:(此处直接按 Enter 键)
Warning:
JKS 密钥库使用专用格式。建议使用"keytool - importkeystore - srckeystore keystore - destkeystore keystore - deststoretype pkcs12"迁移到行业标准格式 PKCS12。
# 将生成的密钥文件拷贝到 azkaban - web 目录
[root@YYR azkaban]# mv keystore azkaban - web
[root@YYR azkaban]#
```

(2) Azkaban Web 服务器配置

AZKABAN WEB 服务器配置

```
# 编辑配置文件
[root@YYR azkaban]# vi azkaban - web/conf/azkaban.properties
# Azkaban Personalization Settings
azkaban.name = My Azkaban
azkaban.label = My Local Azkaban
azkaban.color = #FF3601
azkaban.default.servlet.path = /index
# 该项修改为全路径
web.resource.dir = /usr/local/azkaban/azkaban - web/web/
# 该项修改为上海时间
default.timezone.id = Asia/Shanghai
# Azkaban UserManager class
user.manager.class = azkaban.user.XmlUserManager
# 该项修改为全路径
```

```
user.manager.xml.file=/usr/local/azkaban/azkaban-web/conf/azkaban-users.xml
#Loader for projects
#该项修改为全路径
executor.global.properties=/usr/local/azkaban/azkaban-web/conf/global.properties
azkaban.project.dir=projects
#Velocity dev mode
velocity.dev.mode=false
#Azkaban Jetty server properties.
jetty.use.ssl=false
jetty.maxThreads=25
jetty.port=8081
#该项设置为全路径
jetty.keystore=/usr/local/azkaban/azkaban-web/keystore
#输入密码000000
jetty.password=000000
#Azkaban Executor settings
#mail settings
mail.sender=
mail.host=
#User facing web server configurations used to construct the user facing server URLs. They are useful when there is a reverse proxy between Azkaban web servers and users.
#enduser->myazkabanhost:443->proxy->localhost:8081
#when this parameters set then these parameters are used to generate email links.
#if these parameters are not set then jetty.hostname,and jetty.port(if ssl configured jetty.ssl.port) are used.
#azkaban.webserver.external_hostname=myazkabanhost.com
#azkaban.webserver.external_ssl_port=443
#azkaban.webserver.external_port=8081
job.failure.email=
job.success.email=
lockdown.create.projects=false
cache.directory=cache
#JMX stats
```

```
jetty.connector.stats=true
executor.connector.stats=true
# Azkaban mysql settings by default. Users should configure their own username and password.
database.type=mysql
mysql.port=3306
mysql.host=localhost
mysql.database=azkaban
#修改为MySQL用户名
mysql.user=root
#修改为MySQL用户密码
mysql.password=root00
mysql.numconnections=100
#Multiple Executor
azkaban.use.multiple.executors=true
azkaban.executorselector.filters = StaticRemainingFlowSize,MinimumFreeMemory,CpuStatus
azkaban.executorselector.comparator.NumberOfAssignedFlowComparator=1
azkaban.executorselector.comparator.Memory=1
azkaban.executorselector.comparator.LastDispatched=1
azkaban.executorselector.comparator.CpuUsage=1
```

（3）Azkaban Executor 服务器配置

```
#编辑配置文件
[root@YYR azkaban]# vi azkaban-exec/conf/azkaban.properties

#Azkaban Personalization Settings
azkaban.name=My Azkaban
azkaban.label=My Local Azkaban
azkaban.color=#FF3601
azkaban.default.servlet.path=/index
#该项修改为全路径
web.resource.dir=/usr/local/azkaban/azkaban-web/web/
#该项修改为上海时间
default.timezone.id=Asia/Shanghai
#Azkaban UserManager class
```

AZKABAN EXECUTOR
服务器配置

```
user.manager.class=azkaban.user.XmlUserManager
#该项修改为全路径
user.manager.xml.file=/usr/local/azkaban/azkaban-web/conf/azkaban-users.xml
#Loader for projects
#该项修改为全路径
executor.global.properties=/usr/local/azkaban/azkaban-web/conf/global.properties
azkaban.project.dir=projects
#Velocity dev mode
velocity.dev.mode=false
#Azkaban Jetty server properties.
jetty.use.ssl=false
jetty.maxThreads=25
jetty.port=8081
#Where the Azkaban web server is located
azkaban.webserver.url=http://localhost:8081
#mail settings
mail.sender=
mail.host=
#User facing web server configurations used to construct the user facing server URLs. They are useful when there is a reverse proxy between Azkaban web servers and users.
#enduser->myazkabanhost:443->proxy->localhost:8081
#when this parameters set then these parameters are used to generate email links.
#if these parameters are not set then jetty.hostname,and jetty.port(if ssl configured jetty.ssl.port) are used.
#azkaban.webserver.external_hostname=myazkabanhost.com
#azkaban.webserver.external_ssl_port=443
#azkaban.webserver.external_port=8081
job.failure.email=
job.success.email=
lockdown.create.projects=false
cache.directory=cache
#JMX stats
jetty.connector.stats=true
```

```
executor.connector.stats=true
# Azkaban plugin settings
azkaban.jobtype.plugin.dir=/usr/local/azkaban/azkaban-exec/plugins/jobtypes
# Azkaban mysql settings by default. Users should configure their own username and password.
database.type=mysql
mysql.port=3306
mysql.host=localhost
mysql.database=azkaban
# 修改为 MySQL 用户名
mysql.user=root
# 修改为 MySQL 用户密码
mysql.password=root00
mysql.numconnections=100
# Azkaban Executor settings
executor.maxThreads=50
executor.flow.threads=30
# 新增 executor 端口号配置
executor.port=12321
```

（4）启动 Azkaban

①启动 AzkabanExecutorServer。

```
# 切换当前目录
[root@YYR ~]# cd /usr/local/azkaban/azkaban-exec/bin
# 启动 AzkabanExecutorServer
[root@YYR bin]# ./start-exec.sh
# 查看服务启动状态
[root@YYR bin]# jps
1552 DataNode
1728 SecondaryNameNode
1936 ResourceManager
10533 Jps
10506 AzkabanExecutorServer
1451 NameNode
2044 NodeManager
9452 GradleDaemon
```

启动 AZKABAN 并排错

②激活 AzkabanExecutor。

在浏览器中执行如下命令，激活 AzkabanExecutor：

```
http://192.168.3.190:12321/executor? action=activate
```

③启动 AzkabanWebServer。

```
# 切换当前目录
[root@YYR bin]# cd/usr/local/azkaban/azkaban-web/bin
# 启动 AzkabanWebServer
[root@YYR bin]# ./start-web.sh
# 查看服务启动状态
[root@YYR r bin]# jps
1552 DataNode
1728 SecondaryNameNode
1936 ResourceManager
1451 NameNode
10683 Jps
2044 NodeManager
9452 GradleDaemon
10588 AzkabanExecutorServer
10655 AzkabanWebServer
```

至此，Azkaban 的搭建已经完成，下面检查 Azkaban 运行情况。

任务 3　验证 Azkaban

验证 AZKABAN

在浏览器中输入“http://192.168.3.190:8081”，如果出现如图 7-4 所示界面，表明 Azkaban 启动成功。

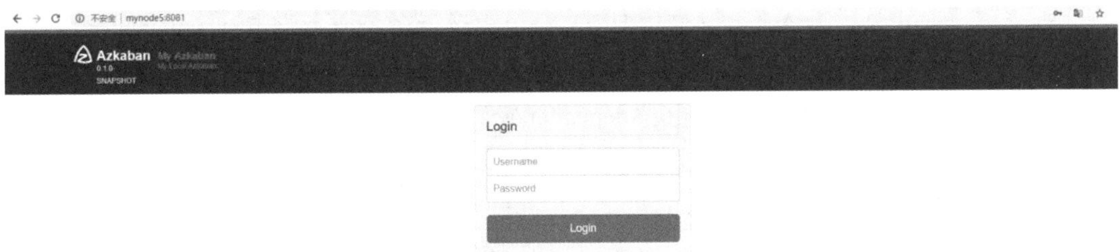

图 7-4　登录界面

默认的用户名和密码都是 Azkaban，可以输入用户名和密码登录 Azkaban 并提交任务流，进行任务管理和调度，如图 7-5 所示。

下面模拟一个任务流来尝试利用 Azkaban 进行任务调度。

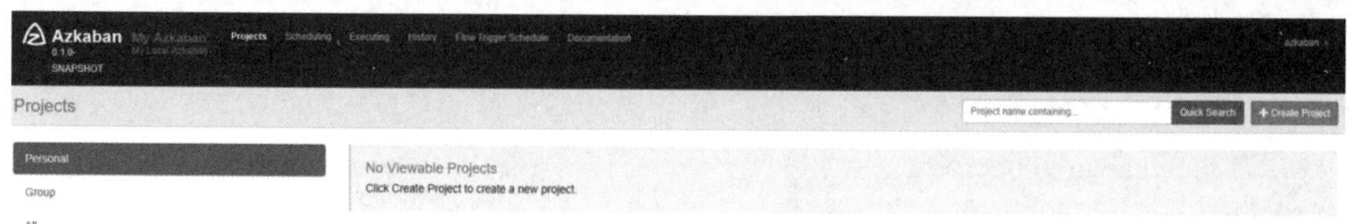

图 7-5　登录 Azkaban

任务 4　构建工作流

构建工作流

本任务中将设计一个任务流 flow，通过这个任务流来学习编写 Azkaban 的任务、在 WebUI 中查看任务流调度及状态。

在 Azkaban 中，一个 project 中可以包含一个或者多个 flow，一个 flow 包含多个 job。这里的 job 是在 Azkaban 中运行的进程，可以是简单的 Linux 命令、Shell 脚本、SQL 脚本等。一个 job 可以依赖于另一个 job，这种多个 job 之间的依赖关系组成 flow，也就是任务流。

1. 设计工作流程

假设现在有 5 个 job，分别是 job1、job2、job3、job4、job5。每个 job 都执行一个 Shell 脚本。job3 依赖于 job1 和 job2 执行的结果，job4 依赖于 job3 执行的结果，job5 依赖于 job4 执行的结果。

按照上述 job 依赖的关系，编写一个简单的任务流提交到 Azkaban 中进行调度执行。

2. 编写各阶段 job

编写 job 非常容易，需要创建一个以“.job”结尾的文本文件，文件的书写格式如下：

```
1.type=command
2.command=需要执行的脚本或命令
```

type=command 是告诉 Azkaban 使用 UNIX 命令去运行命令或者脚本；command=需要执行的脚本或命令就是指定当前 job 需要执行的命令或者脚本。如果当前 job 依赖于其他的 job，只需要在这个文本文件后面加上“dependencies=依赖的 job 名称”即可，依赖的 job 只需写出名称，不需要写出后缀“job”。

为方便演示工作流程，这里设计的每个 job 都调起 Linux 上的一个脚本，脚本中使用简单的 echo 打印一些信息供参考。5 个 job 及对应的脚本设置内容如下：

（1）编写 job 任务

job1.job：

```
1.type=command
2.command=sh/tmp/shellscript/job1.sh
```

job2.job：

```
type=command
command=sh/tmp/shellscript/job2.sh
```

job3. job：

```
type=command
command=sh/tmp/shellscript/job3.sh
dependencies=job1,job2
```

job4. job：

```
type=command
command=sh/tmp/shellscript/job4.sh
dependencies=job3
```

job5. job：

```
type=command
command=sh/tmp/shellscript/job5.sh
dependencies=job4
```

以上 job 任务可以在本地 Windows 环境中编写，编写完成后，需要将 5 个 job 压缩到一个压缩文件中，后期提交到 Azkaban 中执行。

（2）编写脚本内容

job1. sh：

```
echo "开始执行 job1......"
echo "正在执行 job1......"
echo "执行完成 job1......"
```

job2. sh：

```
echo "开始执行 job2......"
echo "正在执行 job2......"
echo "执行完成 job2......"
```

job3. sh：

```
echo "开始执行 job3......"
echo "正在执行 job3......"
echo "执行完成 job3......"
```

job4. sh：

```
echo "开始执行 job4......"
echo "正在执行 job4......"
echo "执行完成 job4......"
```

job5. sh：

```
1.echo "开始执行 job5... ... "
2.echo "正在执行 job5... ... "
3.echo "执行完成 job5... ... "
```

以上脚本是在 Linux 中编写的，保存在/tmp/shellscript/目录（该目录需要创建）下。job1. sh、job2. sh、job3. sh、job4. sh、job5. sh 几个脚本都需要赋予执行权限。

3. 配置工作流并执行操作

在 Azkaban 中任务提交时，必须将所有的 job 文件压缩到一个 zip 文件中，然后再提交到 Azkaban 中执行。首先将以上 5 个 job 压缩到一个 zip 文件中，然后登录 Azkaban，单击右上角的“Create Project”按钮创建一个项目，如图 7－6 所示。

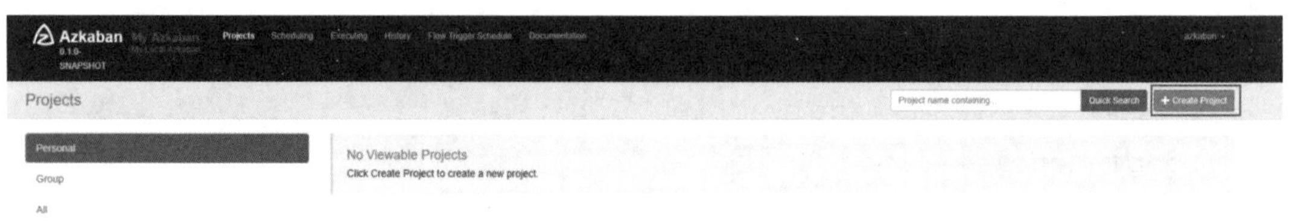

图 7－6　单击“Create Project”按钮

在弹出的框中填写项目名称及项目描述，如图 7－7 所示。

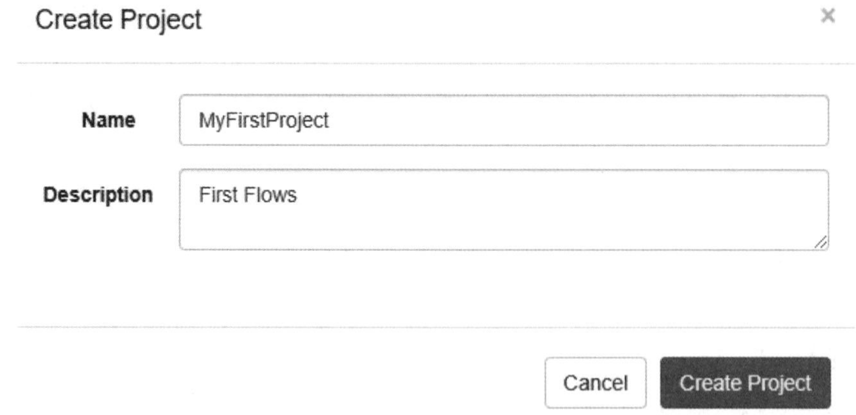

图 7－7　填写项目名称及项目描述

单击“Create Project”按钮，单击“Upload”按钮，上传压缩好的任务，如图 7－8 所示。

上传完成之后，查看任务流，如图 7－9 所示，Azkaban 默认 flow 名称是以最后一个没有依赖的 job 定义的。

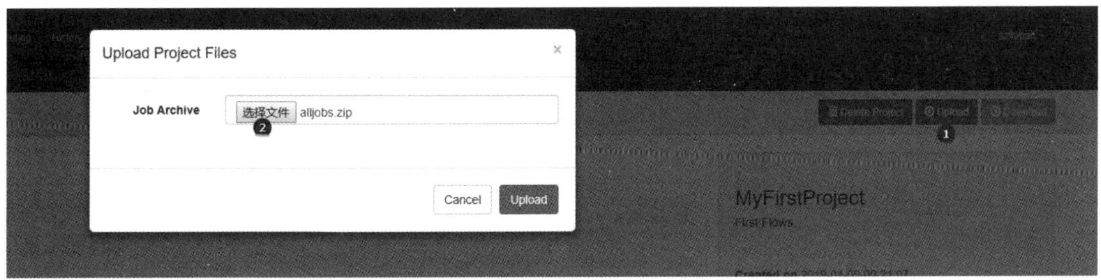

图 7-8　上传任务

Project MyFirstProject

Flows　Permissions　Project Logs

job5　Execute Flow　Executions　Summary

job2

job1

job3

job4

job5

图 7-9　查看任务流

单击“Execute Flow”按钮，可以看到任务流的详细依赖关系，如图 7-10 所示。

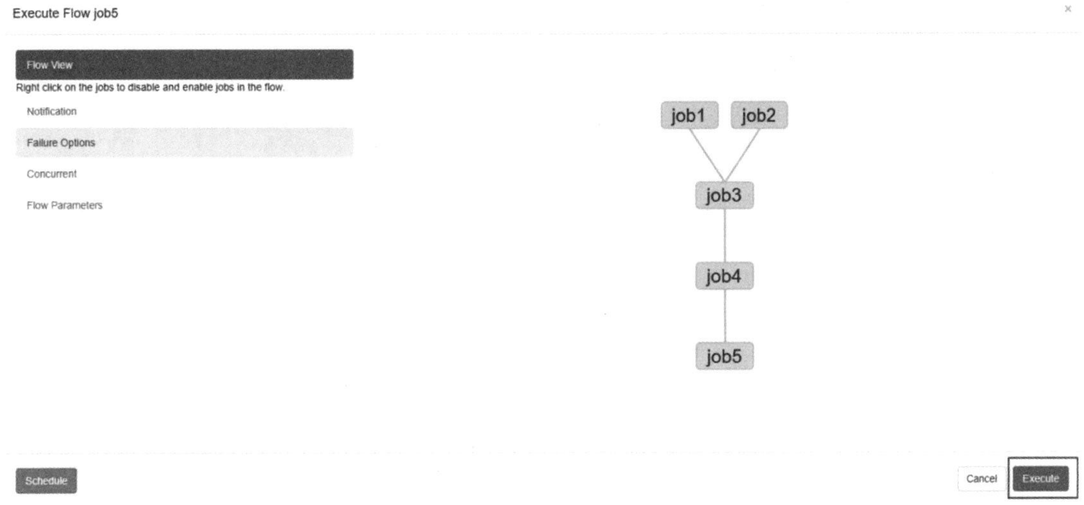

图 7-10　任务流之间关系

单击“Execute”按钮，执行任务。

如果想要每隔一段时间执行一次任务，就单击“Schedule”按钮，配置定时任务，配置好时间之后，单击“Schedule”按钮进行调度即可，如图7－11所示。

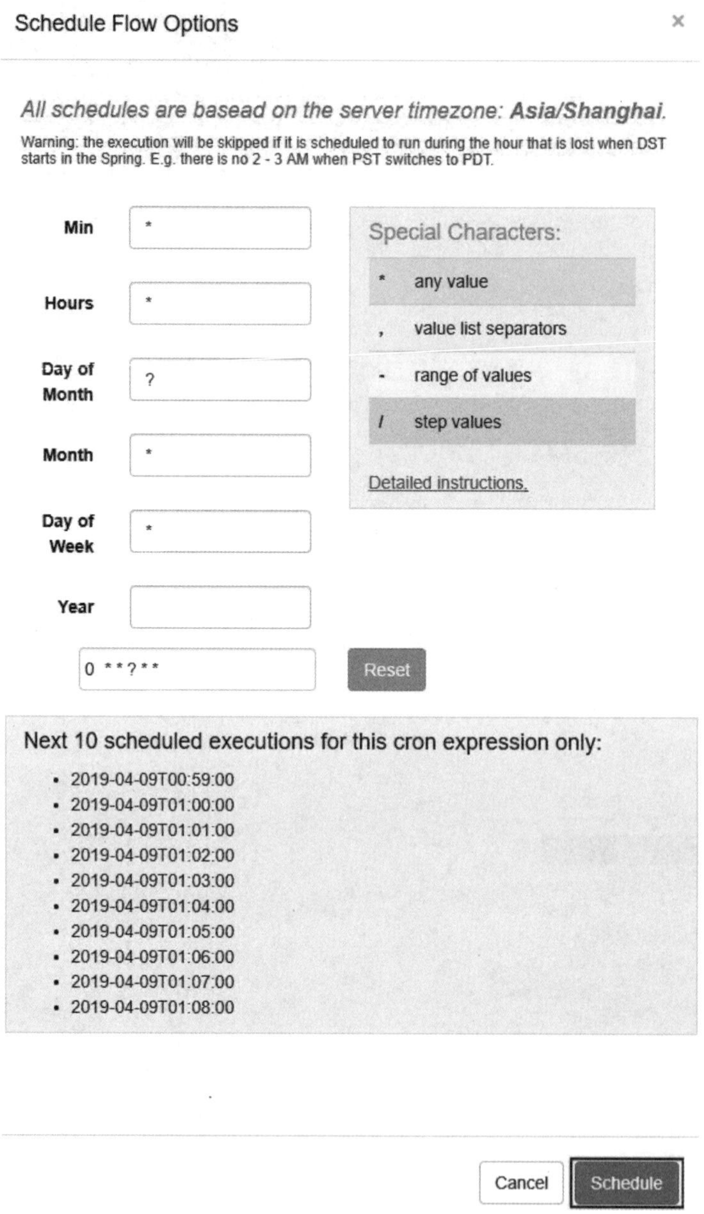

图7－11 Schedule调度

4. 工作流执行监控

任务执行完成之后，会自动跳转到执行成功界面，如图7－12所示。

可以单击“Job List”来查看详细的执行job的先后顺序，如图7－13所示。

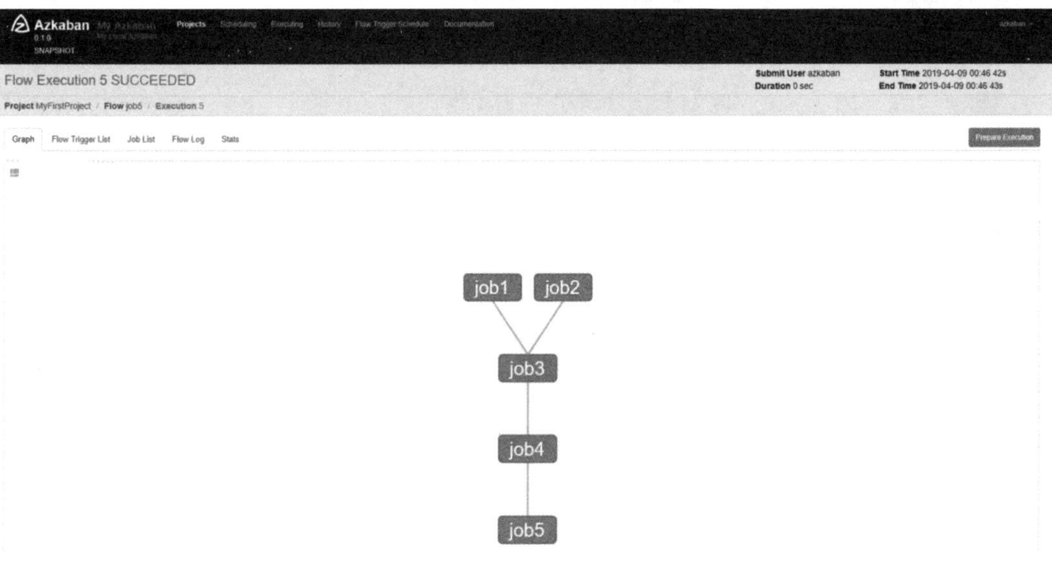

图 7－12　执行成功界面

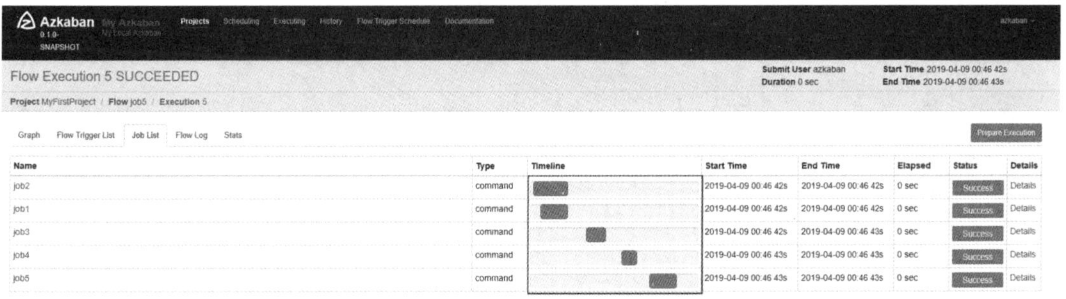

Name	Type	Timeline	Start Time	End Time	Elapsed	Status	Details
job2	command		2019-04-09 00:46 42s	2019-04-09 00:46 42s	0 sec	Success	Details
job1	command		2019-04-09 00:46 42s	2019-04-09 00:46 42s	0 sec	Success	Details
job3	command		2019-04-09 00:46 42s	2019-04-09 00:46 43s	0 sec	Success	Details
job4	command		2019-04-09 00:46 43s	2019-04-09 00:46 43s	0 sec	Success	Details
job5	command		2019-04-09 00:46 43s	2019-04-09 00:46 43s	0 sec	Success	Details

图 7－13　job 执行顺序

还可以通过单击每个 job 的详细信息来查看任务执行过程中产生的日志信息。

例如，单击 job2 的 Details 信息，可以看到：

```
1.09 - 04 - 2020 00:46:42 CST job2 INFO - Starting job job2
  at 1554742002885
2.09 - 04 - 2020 00:46:42 CST job2 INFO - job JVM args: - Dazka-
  ban.flowid = job5 - Dazkaban.execid = 5 - Dazkaban.jobid = job2
3.09 - 04 - 2020 00:46:42 CST job2 INFO - user.to.proxy property was
  not set,defaulting to submit user azkaban
4.09 - 04 - 2020 00:46:42 CST job2 INFO - Building command job execu-
  tor.
5.09 - 04 - 2020 00:46:42 CST job2 INFO - Memory granted for job job2
6.09 - 04 - 2020 00:46:42 CST job2 INFO - 1 commands to execute.
```

```
7. 09-04-2020 00:46:42 CST job2 INFO - cwd = /software/azkaban/az-
   kaban-exec/bin/executions/5
8. 09-04-2020 00:46:42 CST job2 INFO - effective user is:azkaban
9. 09-04-2020 00:46:42 CST job2 INFO - Command: sh/ software/
   test/job2.sh
10. 09-04-2020 00:46:42 CST job2 INFO - Environment variables:{JOB_
    OUTPUT_PROP_FILE = /software/azkaban/azkaban-exec/bin/exe-
    cutions/5/job2_output_6774298763557644376_tmp,JOB_PROP_FILE
    = /software/azkaban/azkaban-exec/bin/executions/5/job2_
    props_595600383798337016_tmp,KRB5CCNAME = /tmp/krb5cc__My-
    FirstProject__job5__job2__5__azkaban,JOB_NAME = job2}
11. 09-04-2020 00:46:42 CST job2 INFO - Working directory:/soft-
    ware/azkaban/azkaban-exec/bin/executions/5
12. 09-04-2020 00:46:42 CST job2 INFO - 开始执行 job2......
13. 09-04-2020 00:46:42 CST job2 INFO - 正在执行 job2......
14. 09-04-2020 00:46:42 CST job2 INFO - 执行完成 job2......
15. 09-04-2020 00:46:42 CST job2 INFO - Process completed success-
    fully in 0 seconds.
16. 09-04-2020 00:46:42 CST job2 INFO - output properties file = /
    software/azkaban/azkaban-exec/bin/executions/5/job2_out-
    put_6774298763557644376_tmp
17. 09-04-2020 00:46:42 CST job2 INFO - Finishing job job2 at
    1554742002940 with status SUCCEEDED
```

至此，Azkaban 环境搭建已完成，经过简单测试，Azkaban 环境能够正常运行。

学习笔记